Contemporary Topics in

POLYMER
SCIENCE

Volume 2

Contemporary Topics in

POLYMER SCIENCE

Volume 2

Edited by

Eli M. Pearce
Polytechnic Institute of New York
Brooklyn, New York

and

John R. Schaefgen
Pioneering Research Laboratory
Textile Fibers Department
E. I. du Pont de Nemours and Company, Inc.
Wilmington, Delaware

PLENUM PRESS · NEW YORK AND LONDON

Library of Congress Cataloging in Publication Data

Main entry under title:

Contemporary topics in polymer science.

Proceedings of the biennial polymer symposia of the A.C.S. Division of Polymer Chemistry held at Key Biscayne, Florida, November 20—24, 1976
Includes indexes.
1. Polymers and polymerization—Congresses. I. Pearce, Eli M. II. Schaefgen, John R. III. American Chemical Society. Division of Polymer Chemistry.

QD380.C64   547'.84   77-21311
ISBN 978-1-4615-6739-4   ISBN 978-1-4615-6737-0 (eBook)
DOI 10.1007/978-1-4615-6737-0

Proceedings of the Eighth Biennial Polymer Symposium of the
Division of Polymer Chemistry held at Key Biscayne,
Florida on November 20—24, 1976

© 1977 Plenum Press, New York
Softcover reprint of the hardcover 1st edition 1977

A Division of Plenum Publishing Corporation
227 West 17th Street, New York, N.Y. 10011

PREFACE

This marks the first publication of the Division of Polymer Chemistry, Inc., American Chemical Society, biennial polymer symposia. This new series will feature new and novel developments in polymer science and technology as presented at these symposia.

This volume reports the proceeding of the Eighth Biennial Polymer Symposium of the Division of Polymer Chemistry held at Key Biscayne, Florida on November 20 - 24, 1976.

It is concerned with a number of developments having both scientific and practical significance. These include polymeric liquid crystal systems in the melt and in solution, polymer blends, and novel polymers in extended chain conformations produced by solid-state polymerization. Rates of conformational transitions and cis-trans isomerizations in polymers, a new look at rubber elasticity theory and new synthetic procedures for stiffening polymer chains are some of the new scientific developments. The book concludes with two approaches for the use of polymers in the important field of slow drug release. The authors are all well-known scientists and include Nobel Laureate Paul Flory, first recipient of the Polymer Division Award which was presented at this Symposium.

Eli M. Pearce
John R. Schaefgen

CONTENTS

THE MOLECULAR THEORY OF RUBBER ELASTICITY

Paul J. Flory

Stanford University

Stanford, California 94305

Once the existence of polymer chains as covalent structures became established a half century ago, the understanding of the molecular basis of the high elasticity characteristic of rubber-like substances presented itself as a foremost challenge. The first attempts[1,2] to explain this remarkable property may be said to date from the beginning of molecular theory as applied to polymers. Although the subject today is old, it is of continuing interest and one in which much remains to be done before an acceptable state of completion will have been attained.

The theory of rubber elasticity is central to polymer science. This property of recoverable, high extensibility is manifested under suitable conditions by virtually all macromolecular substances consisting predominantly of long chains. Moreover, it is exhibited exclusively by materials so constituted. Rubber elasticity is essential to the functioning of elastic proteins and of muscle. It is operative also in the deformation of semi-crystalline polymers, generally not included in the category of rubbers.

The molecular theory of rubber elasticity rests on the premise that the stored elastic free energy, and the force that derives therefrom, originate within the molecular chains comprising the structure, usually a covalent net-

work. The stress comprises the sum of the responses of
the chains to the alteration of their configurations resulting
from the macroscopic strain. Interactions between chains,
although large, are asserted to be of no importance on the
grounds that they are not altered appreciably by the strain
(which orients individual chain units only minutely on the
average).

The validity of this latter assertion , or approximation,
has long been contested. The results of recent experimental investigations on rubber elasticity (cf.seq.),
when compared with theory, provide incontrovertible confirmation of this basic premise. Equally compelling indirect support is provided by recent studies[3] confirming
earlier predictions[4] that the configurations of linear polymer
chains in the amorphous state are random; i.e., they are
unperturbed by interactions with their neighbors, the pro-
fusion of such interactions notwithstanding.[5,6] This being
true, it must follow conversely that interactions with neighbors do not depend appreciably on the chain configuration.
Hence, when the chains undergo orientation by strain,
these interactions do not contribute to the elastic free
energy, and to the stress.

This premise, now securely established, provides us
with a simplification that turns an otherwise vastly complicated situation into one that is tractable. Since it is
legitimate to disregard intermolecular effects, one may
reasonably approach the problem by first considering a
single linear chain.

Elasticity of an Isolated Chain

Chains of the usual length between junctions in a rubber network consist of several hundred skeletal bonds. The
distribution function W(r) for the vector r connecting the
ends of a chain of this length is satisfactorily approximated
by the Gaussian function;[7] i.e.,

$$W(r) = \left(3/2\pi\langle r^2\rangle_0\right)^{3/2} \exp\left[-\left(3/2\langle r^2\rangle_0\right)r^2\right] \quad (1)$$

where $\langle r^2\rangle_0$ is the mean-square separation of the ends of the free chain. It follows that the free energy of the chain is given by

$$A(r) = C - kT \ln W(r)$$

$$= C(T) + \left(3kT/2\langle r^2\rangle_0\right)r^2 \quad (2)$$

where k is Boltzmann's constant; C is a constant and C(T) is a function of temperature. The distribution function W(r) relates directly to the free energy and not to the entropy contribution -TS as assumed in early expositions on rubber elasticity.[4,9-13] Confusion arising from this mistaken identification continues.[14] The average retractive force exhibited by the chain held at fixed length r is

$$\bar{f} = [\partial A(r)/\partial r]_T = 3kT\langle r^2\rangle_0^{-1}r \quad (3)$$

Thus, the average force is directly proportional to r. It is also proportional to the absolute temperature T, and to the inverse of the molecular parameter $\langle r^2\rangle_0$ that characterizes the chain at the chosen temperature T. Proportionality of $\bar{f}$ to r, which underlies the analysis of networks, follows directly from the assertion that W(r) is Gaussian; in fact, eq. 1 may be derived from eq. 3.

Elasticity of Networks

In a network the effects of the macroscopic strain are transmitted to the chains through the junctions at which the ends of chains are multiply joined. The lengths r of the chains are determined by the relative locations of the pairs of junctions joined by linear chains. Hence, the displacements of the junctions by the strain is the issue of pivotal concern.

According to one hypothesis,[4,9-13] displacements ex-
perienced by the junctions are affine (i.e., linear) in the
macroscopic strain. The chain vectors must be deformed
likewise. This hypothesis gains plausibility from consideration of the intertwining of neighboring chains with one
another, which must impede displacements of a junction
relative to its neighbors.[13] Carried to the extreme, this
argument suggests that the junctions be looked upon as
inclusions in a continuum. It therefore lends support to
the stated hypothesis. Hence, the average (denoted by
$\langle \ \rangle$) of r^2 over all chains of the network is

$$\langle r^2 \rangle = \langle x^2 \rangle + \langle y^2 \rangle + \langle z^2 \rangle$$

$$= \left(\lambda_x^2 + \lambda_y^2 + \lambda_z^2 \right) \langle r^2 \rangle_0 / 3 \tag{4}$$

where λ_x, λ_y, λ_z are the principal extension ratios that
specify the strain relative to an isotropic state of reference
of volume V_0 such that $\langle r^2 \rangle = \langle r^2 \rangle_0$. That is,

$$\lambda_x^2 = \langle x^2 \rangle / \langle x^2 \rangle_0 = 3 \langle x^2 \rangle / \langle r^2 \rangle_0$$

etc., where by definition

$$\lambda_x = L_x / L_{x0} \tag{5}$$

etc., L_x and L_{x0} being dimensions of the sample when deformed and when in the reference state, respectively.

On the basis of the stated hypothesis, the elastic free
energy of the network comprises two terms:[4] one for
the changes in configurations within the chains, and the
other due to the dispersion of the junctions, μ in number,
over the volume V. The former term, being the sum of the
contributions of ν individual chains, is obtained through
multiplication of eq. 2 by ν and replacement of r^2 with
the average $\langle r^2 \rangle$ over all chains of the network. The latter
term is the gas-like expression $-\mu kT \ell n(V/V^0)$. Combining
these terms, and replacing $\langle r^2 \rangle$ according to eq. 4, one

obtains[4,15]

$$\Delta A_{el} = \frac{1}{2} \nu kT\left(\lambda_x^2 + \lambda_y^2 + \lambda_z^2 - 3\right) - \mu kT \ell n(V/V_0) \qquad (6)$$

for the free energy relative to the state of reference specified above. For a tetrafunctional network, $\mu = \nu/2$.

The retractive force in simple elongation is given by $f = \left(\partial \Delta A_{el}/\partial L\right)_{T,V}$ where $L = \lambda L_0$ is the length in the direction of stretch. The extension ratio in each of the transverse directions is $(V/V_0\lambda)^{1/2}$.

$$f = \left(\nu kT/L_0\right)\left[\lambda - (V/V_0)\lambda^{-2}\right] \qquad (7)$$

or

$$f = \frac{\nu kT}{L_0}\left(\frac{V}{V_0}\right)^{1/3}(\alpha - \alpha^{-2}) \qquad (8)$$

where $\alpha = L/L_i$ is the elongation relative to the isotropic sample at the final volume V, where its length is L_i. Thus, $\alpha = (V/V_0)^{1/3}\lambda$ and $L_i = (V/V_0)^{1/3}L_0$.

In the model of James and Guth[16] each chain of the network imparts a force between the junctions so connected that is proportional to the distance between them (see eq.3) but the chain is devoid of all other material properties. The chains may pass through one another freely and parts of two or more of them may occupy the same space. Being free of constraints by neighboring chains, the junctions of this "phantom network"[17] may undergo displacements that are independent of their immediate surroundings.

James and Guth[16] showed that (i) the mean positions of the junctions in such a network are linear (i.e., affine) in the macroscopic strain, and (ii) fluctuations about these mean positions are independent of the strain. Hence, the

mean values of the components, $\bar{x}_{ij}$, $\bar{y}_{ij}$, and $\bar{z}_{ij}$, of the
chain vector r_{ij} joining junctions i and j are proportional
to λ_x, λ_y and λ_z, respectively; fluctuations Δr_{ij} from $\bar{r}_{ij}$ are
independent of the strain. The instantaneous chain vector
is $r_{ij} = \bar{r}_{ij} + \Delta r_{ij}$. The time-average of the squared magni-
tude of r_{ij} is $\langle r_{ij}^2 \rangle = (\bar{r}_{ij})^2 + \langle (\Delta r_{ij})^2 \rangle$. For a "perfect"
tetrafunctional network[4] in its reference state, where
$\lambda_x = \lambda_y = \lambda_z = 1$, it may be shown[17-19] that

$$\bar{r}_0^2 = \langle (\Delta r)^2 \rangle = \frac{1}{2} \langle r^2 \rangle_0 \qquad (9)$$

Thus, only half of $\langle r^2 \rangle_0$ is subject to alteration by deformation of the network. The other half, $\langle (\Delta r)^2 \rangle$, representing
the fluctuations, is invariant with deformation. In conse-
quence of this circumstance, the elastic free energy for a
perfect tetrafunctional network is half of the first term of
eq. 6; i.e.,[16-20]

$$\Delta A_{el} = \frac{1}{4} \nu kT \left(\lambda_x^2 + \lambda_y^2 + \lambda_z^2 - 3 \right) \qquad (10)$$

A term in $\ln(V/V^0)$ does not occur owing to the invariance of
the dispersion of the junctions about their mean positions,
fully defined by the strain.[17] Generalization to a network
of any functionality, and any degree of imperfection, is accomplished[17] by replacing $\nu/4$ by $\xi/2$ where ξ is the cycle
rank of the network; i.e., ξ is the number of cuts required
to reduce the network to a "tree" in which all cyclic paths
are eliminated without fragmenting it into separate "graphs,"
or structures.

Stress-strain relations derived from eq. 10 are of the
same form as eqs. 7 and 8 derived from eq. 6. They differ
only by a numerical factor arising from replacement of ν
by ξ. For a tetrafunctional network this factor is one-half.
Hence, the relationship of the retractive force to the extension for simple elongation of a tetrafunctional phantom
network is given by eq. 7 or eq. 8 modified by this
factor.[16-20] The factor depends on the functionality of the
network.[17]

Comparisons with Experiment

The form of the stress-strain curve for rubbers subjected to unidirectional extension shows significant departures from theory. This is illustrated in Fig. 1 where curves calculated from eq. 8, shown dashed, are compared with the solid curve calculated according to the Mooney-Rivlin[21,22] empirical equation which can be written

$$f = 2C_1'(V/V_0)^{1/3} \left(1 + C_2'/C_1'\alpha\right)\left(\alpha - \alpha^{-2}\right) \qquad (11)$$

Here C_1' and C_2' are empirical constants; the term $C_2'/C_1'\alpha$ measures the departure from theory according to eq. 7 or 8.

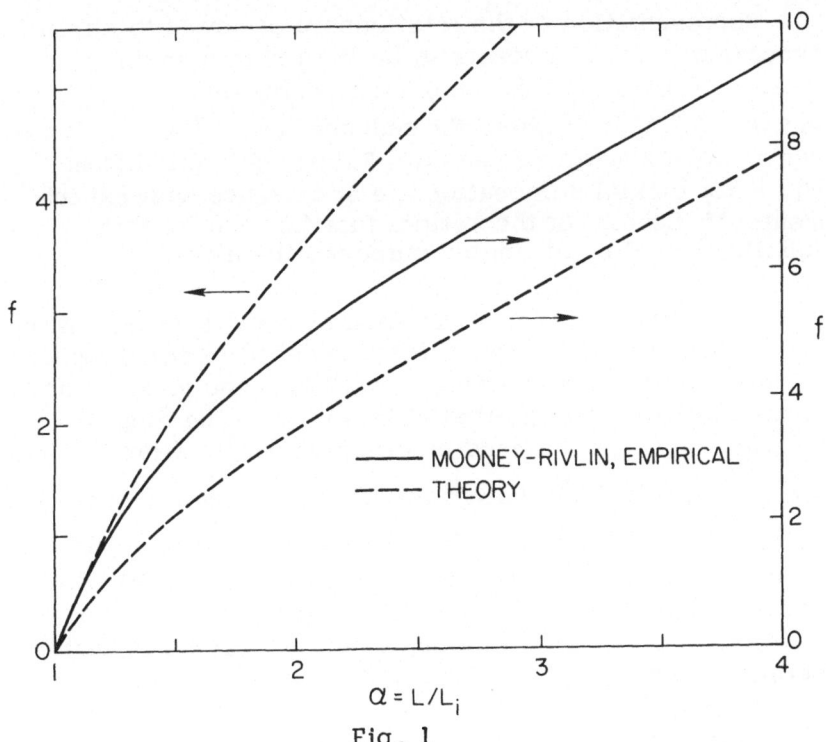

$$\alpha = L/L_i$$

Fig. 1

Eq. 11 reproduces experimental results for simple extension $(\alpha > 1)$ fairly well.[14,23-25] The behavior of virtually all types of rubbers is remarkably similar in this respect. The observed force plotted against elongation typically displays greater curvature than the theory predicts. The two theoretical curves shown are in fact the same curve plotted on different ordinates, as is indicated. The one is scaled to match the initial slope of the Mooney-Rivlin curve, and the other converges to the ultimate slope of that curve as α increases indefinitely.

The same curves are shown in Fig. 2 where the reduced force $f/(\alpha - \alpha^{-2})$ is plotted against $1/\alpha$, as eq. 11 suggests. The range ordinarily covered by experiments is heavy-lined. The deviations from theory are large. They are invariably decreased by swelling.[14,23-27]

Representation of the relationship of stress to strain according to eq. 11 appears to be limited to uniaxial elongation. It fails utterly for uniaxial compression $(\alpha < 1)$,[14,25,28,29] where the reduced force $f/(\alpha - \alpha^{-2})$ becomes approximately constant. Relations inferred from eq. 11 for biaxial deformation are at variance with experiments.[14] Search for theoretical justification for this equation would be of limited purpose, therefore.

In an exhaustive investigation of natural rubber, Allen and co-workers[24] showed that the reduced forces determined at various dilutions covering a fourfold range yielded the same intercept when plotted as in Fig. 2. The slopes, reflecting C_2'/C_1', decreased progressively with dilution, but the intercepts, $2C_1'$, showed no discernible dependence on dilution. This conclusion has been confirmed for poly (dimethylsiloxane)[27] and for cross-linked polyisobutylene.[30] These observations indicate that the intercept is a fundamental characteristic of the material. It may be identified, perhaps, with the molecular quantity $\nu kT/L_0$ appearing in eq. 8, or with $\xi kT/L_0$ in the theory for phantom networks. Measurements carried out on networks so prepared as to

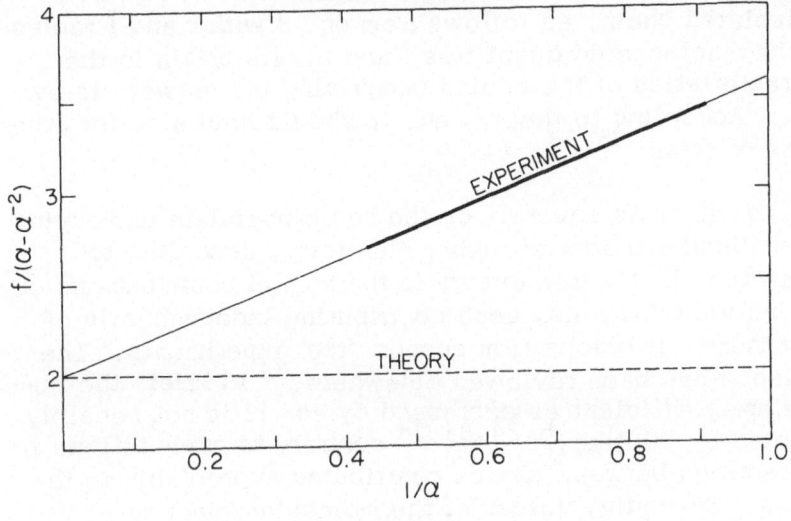

Fig. 2

permit estimation of ν and of ξ from chemical evidence support identification with the latter quantity.[31,32]

Studies on the change of stress with temperature at fixed deformation have yielded results of the foremost significance. Consider the dependence of f/T on the temperature T at fixed volume V and length L. Then $\alpha = L/L_i$ is fixed also since L_i is determined by V (see above). According to eq. 8, therefore, f/T depends on $\left(V_0^{1/3} L_0 \right)^{-1}$ under the stated conditions. The length L_0 in the reference state is proportional to $V_0^{1/3}$ and, according to the definition of the reference state, V_0 is the volume that renders $\langle r^2 \rangle = \langle r^2 \rangle_0$ at the given temperature. Hence, V_0 is proportional to $\langle r^2 \rangle_0^{3/2}$, from which it follows that[8,15]

$$[\partial \ln(f/T)/\partial T]_{V,L} = - d\ln \langle r^2 \rangle_0/dT \qquad (12)$$

The same relation obviously holds for a phantom network, for which the equation of state differs from eq. 8, or eq. 7, only by a numerical factor. Eq. 12 is applicable also to an isolated chain, as follows from eq. 3 with r and $\bar{f}$ replaced by the macroscopic quantities f and L. Its origin in the characteristics of the chains comprising the network is evident. According to theory, eq. 12 should hold also for other types of strain.[15,33]

Eq. 12 rests squarely on the basic postulate underlying the molecular theory of rubber elasticity, according to which the elastic free energy is the sum of contributions of the individual chains, each contributing independently of the others. It enjoys firm support from experiments. The evidence has been reviewed elsewhere.[34] In brief, the temperature coefficient as expressed by eq. 12 is not sensibly affected by dilution,[6,24,34,35] contrary to expectations if interactions between chains contributed appreciably to the stress. Secondly, values of the molecular quantity $d\ln\langle r^2\rangle_0/dT$ derived from experimental measurements of f as a function of T for a macroscopic network are in good agreement with the best values of the former quantity obtained from measurements carried out on dilute solutions of the corresponding linear polymer.[6,34,36]

That the stress originates within the chains comprising the network is well established by the evidence cited. The conclusion receives indirect, but equally compelling, support from the investigations on the configurations of chains in the bulk polymer, as pointed out above.

This unequivocal confirmation of the basic premise stands in sharp contrast to the inadequacy of theory in accounting accurately and comprehensively for the relationship of stress to strain. In this circumstance, critical reexamination of network topology and of the displacements of junctions under strain seemed to be indicated.

Structure and Topology of Real Network

As may be shown by simple calculations, the junctions directly connected by chains to a given junction are not its nearest spatial neighbors in a typical elastomeric network.[17]

Within the domain defined crudely by the locations of these closest topological neighbors (four in number in a tetra-functional network) as many as fifty or more other junctions may reside. The situation is depicted in Fig. 3. The paths (not shown) that connect spatial neighbors, indicated by X in the figure, to the central junction may include many chains; we shall be content to let these connections remain obscure. It suffices to observe that elastomeric networks are copiously interpenetrated structures, as these considerations clearly show. They bear little resemblance to a lattice- or even to a disordered lattice - often invoked incorrectly as a suitable analog.

Fig. 3 may be considered to represent the instantaneous positions of junctions of a phantom network. Since the average values $\bar{r}$ of the chain vectors r are affine in the strain but the fluctuations are not, the junctions that at a given instant occupy some small region of the space will undergo different displacements under strain. Hence, the

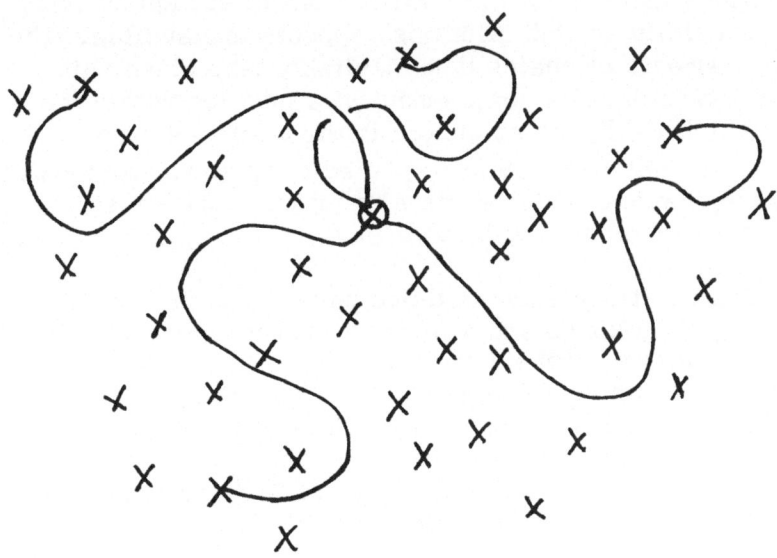

Fig. 3

set of junctions that, according to their mean positions, are eligible to occur as closest neighbors to a selected junction, will not be conserved under strain; the mean positions of some of them will be displaced beyond the range of fluctuations, and, simultaneously, others may be brought closer.[17]

In view of the intricate involvements of junctions and their pendant chains in a real network, full compliance with the characteristics of the phantom model network can scarcely be expected. Fluctuations of the junctions about their mean positions may be severely impeded. More important, relocation, under strain, of the neighbors about a given junction may be difficult; cooperative rearrangement of chains obviously is required. These concerns lend support to the view of W. Kuhn[13] that the junctions of a network are firmly embedded in the matrix provided by neighbors. Fluctuations deduced for a phantom network are frozen in, and the junctions must undergo displacements that are affine in the macroscopic strain, according to this view. Eqs. 6-8 would then hold without alteration.

Strict adherence to this "affine" model at high strains, and especially at high dilutions, appears implausible. The rearrangements of chains that inevitably take place with strain must relax the confinement of a junction within the domain defined by its neighbors in the reference state. Hence, the network may be expected to approach conformity with the phantom model as its elongation is increased, and especially when dilated by swelling.

Considerations such as these have led to the hypothesis put forward by Ronca and Allegra,[37] and independently by the author,[17] that the behavior of real networks is a compromise between the two models considered. Unswollen networks at small deformations may adhere closely to the first model. With increase in elongation they may be expected to approach the phantom network in their behavior. It is as if the upper dashed curve in Fig. 1 is preferred initially at small strains, but as the elongation increases the force tends toward the lower dashed curve. This hypothesis immediately explains

the failure of either theory to account for the form of the
observed stress-strain curve; the respective theories predict curves of the same form but with different coefficients
of proportionality. Transition from one to the other with
elongation would account for the departure of experiment
from both.

Toward an Improved Theory

In an effort to give quantitative expression to the hypothesis described above, a model has been devised that is
susceptible to mathematical treatment.[38] It is illustrated
in Fig. 4. Here $\bar{R}_i$ denotes the mean position for junction i
in the given state of strain if it was subject only to the

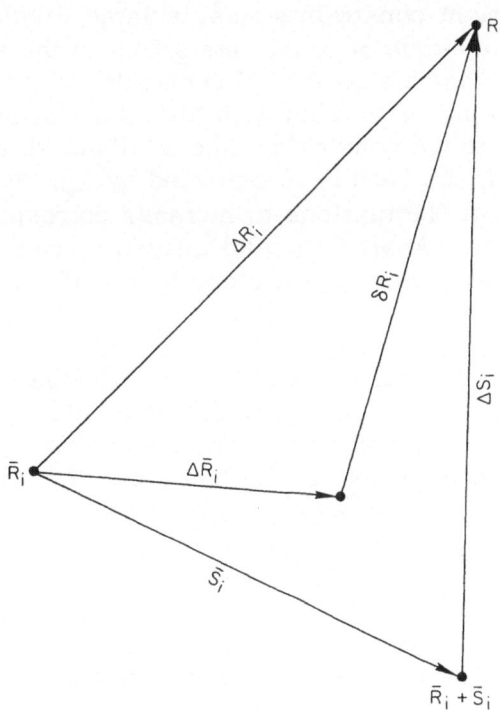

Fig. 4

influences of the chains attached to it, i.e., if phantom behavior prevailed. The fluctuation from this mean position is $\Delta\underset{\sim i}{R}$. We suppose that, during one of the excursions of this junction of the phantom network, it becomes ensnarled by neighboring chains suddenly endowed with material properties, so that it is henceforth subject to "entanglement" constraints operating from a point located at $\bar{\underset{\sim i}{R}} + \bar{\underset{\sim i}{s}}$. If the junction subsequently occurs, momentarily, at $\bar{\underset{\sim i}{R}} + \Delta\underset{\sim i}{R}$, it is acted upon by two forces, one generated by the response within the network to the displacement $\Delta\underset{\sim i}{R}$ from the (phantom) mean position $\bar{\underset{\sim i}{R}}$ and the other from its displacement by $\Delta\underset{\sim i}{s}$ from the center of action of the entanglement constraints to which it is subject. Under the influence of both constraints, its mean position is displaced by $\Delta\bar{\underset{\sim i}{R}}$ from $\bar{\underset{\sim i}{R}}$.

When the specimen is deformed, the centers of the entanglement constraints must undergo displacements which, on the average at least, are affine in the strain. We assume further that the domain of constraint is likewise deformed by the strain, in keeping with the argument above that the action of these constraints should diminish at high strains. Thus, if the sample is extended by λ_x, we assume the tolerance of fluctuations to increase correspondingly in this direction. Apart from qualitative arguments in its support, this assumption is warranted by expediency rather than rigor.

The enhancement of the elastic free energy in consequence of the constraints due to neighboring chains comprises two terms. The first arises from alterations of the displacements $\Delta\underset{\sim i}{R}$ of the junctions from their distribution for the phantom network. It represents the free energy associated with the response of the network to displacements of its junctions from their mean positions in absence of the constraints by neighbors. The second term arises similarly from the displacements $\Delta\underset{\sim i}{s}$ of the junctions from their mean positions if the constraints by neighbors alone were operative. From the combination of these terms the contribution to the stress attributable to "entanglements" may be obtained.

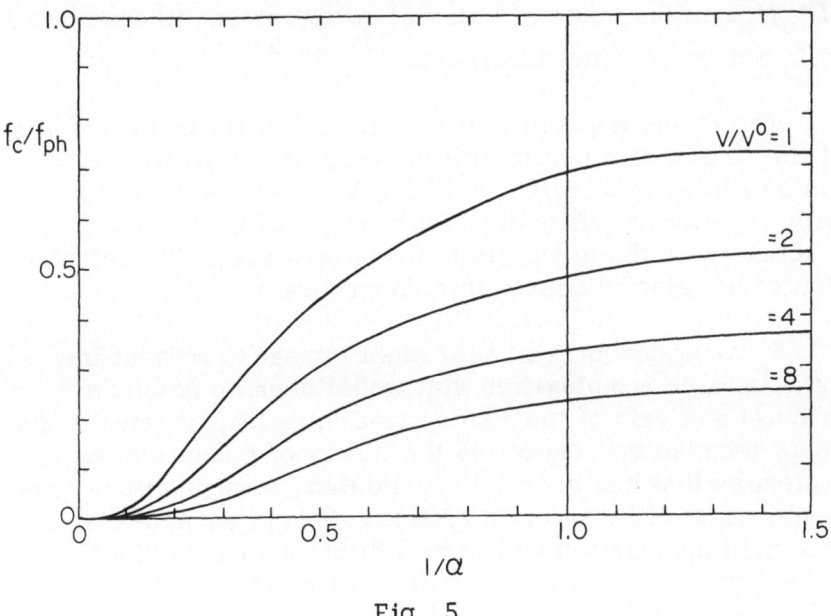

Fig. 5

In Fig. 5 the contribution f_c of these constraints to the tension in simple elongation, expressed relative to the force f_{ph} for the phantom network, is plotted against $1/\alpha$. The curves have been calculated for an arbitrary relative severity of the entanglement constraints ($\kappa = 10$; see ref. 38) and for the dilutions indicated by the ratios V/V_0. If the intercept at $1/\alpha = 0$ in Fig. 2 is identified with the reduced force for the phantom network, then the calculated ratios f_c/f_{ph} may be compared with the difference between the reduced force (solid line) in Fig. 2 and the intercept at $1/\alpha = 0$ (dashed line), this difference being divided by the intercept.

The calculated relation is nonlinear in the range $1/\alpha < 1$. The magnitude of f_c/f_{ph} at $\alpha \approx 1$ compares favorably with the enhancement of the reduced force observed in experiments[25] and conventionally attributed to the term in C_2, the origin of which has long been a mystery. With dilution, f_c/f_{ph} decreases as is observed. Finally, the approximate constancy of f_c/f_{ph} in the range of uniaxial compression, where $1/\alpha > 1$, is in agreement with experiment.[14,25,29]

The theory appears to reproduce all of the main features of the stress-strain curve for uniaxial deformation. Only the nonlinearity of f_c/f_{ph} with $1/\alpha$ in extension is at variance with experiment. This deficiency may reflect quantitative inaccuracy in the assumption above concerning the deformation of domains of constraints under strain.

Polymer networks at first sight appear to present insurmountable complexities that would seem to preclude rational analysis of their properties in molecular terms. The basic premise that underlies the theory of rubber elasticity, a premise that has been fully validated, permits circumvention of most of these complexities. Recent advances of theory in conjunction with a wealth of empirical evidence gained from well chosen, carefully executed experiments offer the prospect of a comprehensive understanding of the elastic equation of state and associated properties of elastomeric materials in the foreseeable future.

References

1. K. H. Meyer, G. vonSusich and E. Valkó, Kolloid Z., 59, 208 (1932).
2. E. Guth and H. Mark, Monatsh. Chem., 65, 93 (1934).
3. See for example, J. P. Cotton et al., Macromolecules, 7, 863-872 (1974); R. G. Kirste, W. A. Kruse, and K. Ibel, Polymer, 16, 120-124 (1975); J. Schelten, G. D. Wignall and D. G. H. Ballard, Polymer, 15, 682 (1974); Eur. Polymer J., 10, 861 (1974); E. W. Fischer, G. Leiser and K. Ibel, Polymer Lett., 13, 39 (1975).
4. P. J. Flory, "Principles of Polymer Chemistry," Cornell Univ. Press, Ithaca, N. Y., 1953.
5. P. J. Flory, J. Macromol. Sci., Phys. Ed., B12, 1-11 (1976).

6. P. J. Flory, Pure & Appl. Chem., Macromolecular Chem., <u>8</u>, 1-15 (1972); or, Rubber Chem. and Tech. <u>48</u>, No. 3, 513-525 (1975).

7. P. J. Flory and D. Y. Yoon, J. Chem. Phys.,<u>61</u>,5358-5365 (1974). P. J. Flory and V. W. C. Chang, Macromolecules, <u>9</u>, 33-40 (1976).

8. M. V. Volkenstein and O. B. Ptitsyn, Zh. Tekh. Fiz., <u>25</u>, 662 (1955). M. V. Volkenstein, "Configurational Statistics of Polymeric Chains" (translated from the Russian edition by S. N. Timasheff and M. J. Timasheff), Interscience, New York, 1963, pp. 501-7.

9. E. Guth and H. M. James, Ind. Eng. Chem., <u>33</u>, 624 (1941). H. M. James and E. Guth, J. Chem. Phys.,<u>11</u>, 455 (1943).

10. F. T. Wall, J. Chem. Phys., <u>11</u>, 527 (1943).

11. L. R. G. Treloar, Trans. Faraday Soc., <u>39</u>, 36, 241 (1943).

12. P. J. Flory and J. Rehner, Jr., J. Chem. Phys., <u>11</u>, 512 (1943).

13. W. Kuhn, J. Polymer Sci., <u>1</u>, 380 (1946).

14. L. R. G. Treloar, "The Physics of Rubber Elasticity," 3rd ed., Oxford Univ. Press, 1975.

15. P. J. Flory, Trans. Faraday Soc., <u>57</u>, 829 (1961); J. Am. Chem. Soc., <u>78</u>, 5222 (1956).

16. H. M. James, J. Chem. Phys., <u>15</u>, 651 (1947). H. M. James and E. Guth, ibid., <u>15</u>, 669 (1947).

17. P. J. Flory, Proc. Roy Soc., in press. Presented at the Meeting for Discussion of Rubber Elasticity held by the Royal Society in London, 20 Nov., 1975.

18. B. E. Eichinger, Macromolecules, <u>5</u>, 496 (1972).

19. W. W. Graessley, Macromolecules, <u>8</u>, 865 (1975).

20. J. A. Duiser and J. A. Staverman, "Physics of Non-Crystalline Solids, J. A. Prins, Ed., North Holland Publishing Co., Amsterdam (1965), pp. 376-387.

21. M. Mooney, J. Appl. Phys., <u>11</u>, 582 (1940.).

22. R. S. Rivlin, Phil. Trans. Roy. Soc. (London), <u>A241</u>, 379 (1948).

23. S. M. Gumbrell, L. Mullins and R. S. Rivlin, Trans. Faraday Soc., <u>49</u>, 1495 (1953).

24. G. Allen, M. J. Kirkham, J. Padget and C. Price, Trans. Faraday Soc., 67, 1278-1292 (1971).
25. J. E. Mark, Rubber Chem. Technol., 48, 495 (1975).
26. D. S. Chiu and J. E. Mark, Colloid and Polymer Science, in press.
27. P. J. Flory and Y. Tatara, J. Polymer Sci.: Polymer Phys. Ed., 13, 683-702 (1975).
28. L. R. G. Treloar, Rep. Prog. Physics, 36, 755 (1973).
29. R. Y. S. Chen, C. U. Yu and J. E. Mark, Macromolecules, 6, 746 (1973). C. U. Yu and J. E. Mark, Polymer J., 7, 101 (1975).
30. H. Neidlinger and P. J. Flory, unpublished.
31. D. J. Walsh, G. Allen and G. Ballard, Polymer, 15, 366 (1974). G. Allen, P. A. Holmes and D. J. Walsh, Faraday Disc. Chem. Soc., 57, 19 (1974). G. Allen, P. L. Egerton and D. J. Walsh, Polymer, 17, 65 (1976).
32. J. E. Mark, J. Chem. Phys., in press.
33. G. Allen, C. Price and N. Yoshimura, J. Chem. Soc., Faraday Trans. I, 71, 748 (1975).
34. J. E. Mark, Rubber Chem. Technol., 46, 593 (1973).
35. A. Ciferri, C. A. J. Hoeve and P. J. Flory, J. Amer. Chem. Soc., 83, 1015 (1961).
36. P. J. Flory, A. Ciferri, and R. Chiang, J. Amer. Chem. Soc., 83, 1023 (1961). R. Chiang, J. Phys. Chem., 70, 2348 (1966).
37. G. Ronca and G. Allegra, J. Chem. Phys., 63, 4990 (1975).
38. P. J. Flory, to be published.

LIQUID CRYSTALLINE SOLUTIONS FROM POLYHYDRAZIDES
IN AQUEOUS ORGANIC BASES

J. D. Hartzler and P. W. Morgan

E. I. du Pont de Nemours & Co., Inc.

Experimental Station, Wilmington, Delaware

Aromatic polyhydrazides are well known from
the work of Frazer, Wallenberger and Sweeny (1,2)
as precursors to poly-1,3,4-oxadiazoles. Polyamide-hydrazides have been described in detail by
Black, Preston, and coworkers (3,4,5) and by
Culbertson and Murphy (6). High tenacity, high
modulus fibers have been made from poly(terephthalic
hydrazide)(7) and from polyamide-hydrazides with
ordered structures (5).

Throughout these studies there is no indication
that any of the polymers have ever yielded liquid
crystalline solutions. In fact, one reference
reports an unsuccessful attempt to form such a
solution in dimethyl sulfoxide-LiCl (8) and another
article (9) explains that liquid crystalline
solutions may not be formed because of rotational
freedom about the NH-NH bond of the hydrazide units.

The results reported here show that polyhydrazides do yield liquid crystalline solutions
in a variety of solvents and, therefore, can
exist as extended chain structures in solution.

RESULTS AND DISCUSSION

Polymer Preparation

Polyhydrazides were prepared by the general
procedure of Frazer (1) from dihydrazides and
diacid chlorides in amide solvents (Fig. 1).

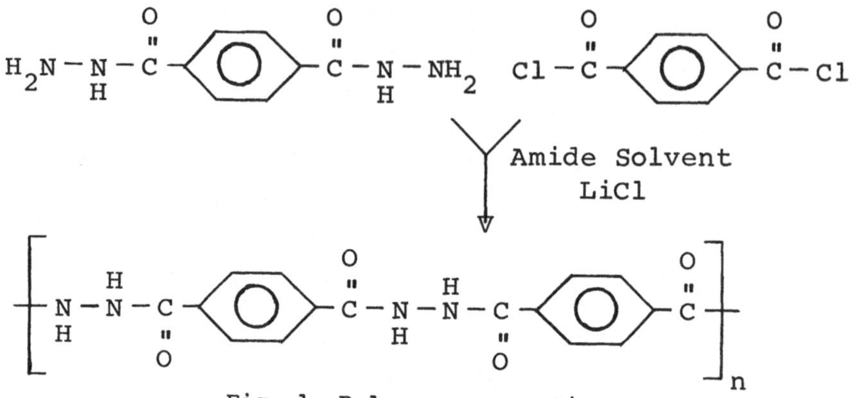

Fig. 1. Polymer preparation.

Some of the hydrazides have low solubility in the
amide solvents and, therefore, slow addition of
the acid chlorides is in order to minimize side
reactions and permit the hydrazide monomers to
dissolve as they react. The degree of polymerization is also enhanced by the addition of lithium
chloride to the solvent. An optimum amount is
about one mole of lithium chloride/mole of amide
group in the formed polymer (Fig. 2). Larger
amounts depressed the product inherent viscosity.
Mixtures of hexamethylphosphoramide with dimethylacetamide or N-methylpyrrolidone have been found
to yield higher molecular weights than the single
solvents in the preparation of some wholly aromatic

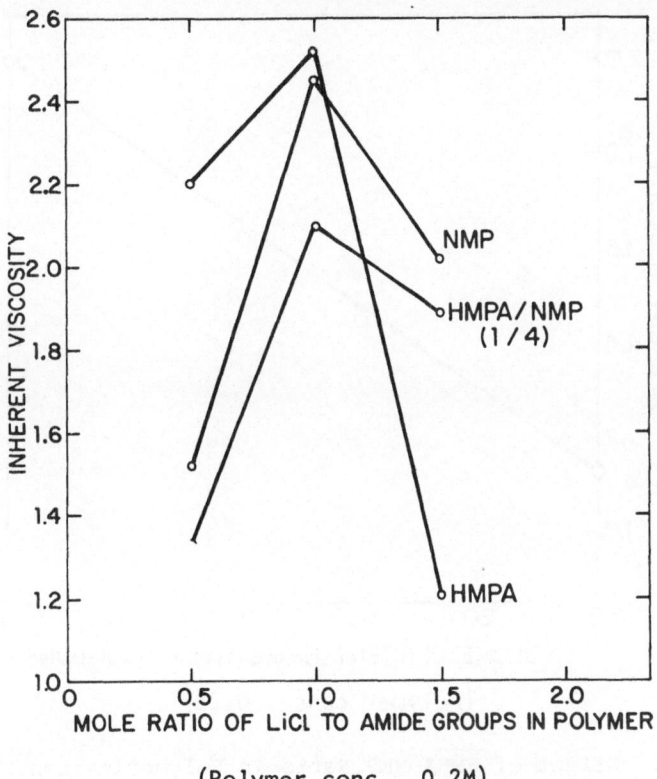

(Polymer conc., 0.2M)

Fig. 2.   Inherent Viscosity of O-T-O-CIT as a Function of
LiCl Concentration in Polymerization Medium

polyamides, such as poly(p-phenylene terephthal-
amide).   There was no such synergistic effect in
the preparation of poly(terephthalic-co-chloro-
terephthalic hydrazide) in hexamethylphosphoramide*-
N-methylpyrrolidone with added lithium chloride
(Fig. 3).

---

* Hexamethylphosphoramide has been found in
  laboratory experiments to be carcinogenic in
  rats.

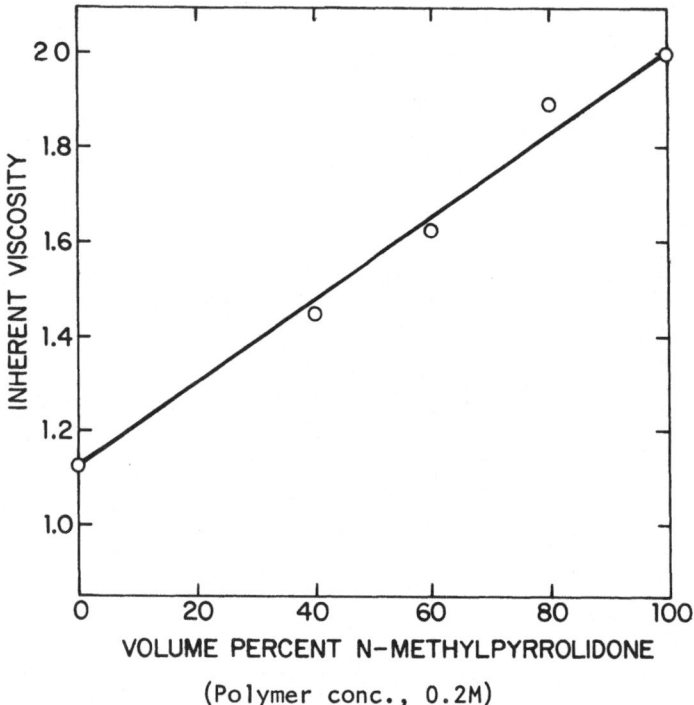

(Polymer conc., 0.2M)

Fig. 3. Effect of HMPA-NMP Ratio in Polymerization Medium
on Inherent Viscosity of O-T-O-CIT (LiCl Content, 5g/100 ml
Solvent)

All of the polyhydrazides can be isolated as
essentially colorless granular powders.
Occasionally samples are obtained with an
appreciable yellow coloration. This is due to
the presence of residual base and can be removed
by washing the freshly formed polymer with dilute
acid.

Polyhydrazide Solutions

Polymer Solubility in Aqueous Organic Bases.
O-T [poly(terephthalic hydrazide)] and many other

polyhydrazides are soluble in a variety of aqueous organic bases which have pKa values greater than about 10.4 (10). Primary and tertiary bases were poor solvents for O-T with maximum solubility less than 5%. Examples are 5% aqueous tertiary-butyl-amine and 5% aqueous triethylamine. Secondary amines were better solvents, with solubility reaching ~9% in diethylamine and piperidine with the formation of liquid crystalline (anisotropic) solutions (Table I). Other useful secondary amines were methylbutylamine (5-15%) and pyrrolidone (15%), but these yielded only isotropic solutions. Solubility in aqueous quaternary ammonium bases increased as the size of the organic radicals decreased; that is, solubility decreased in the order, methyl > ethyl > propyl > butyl. In this

TABLE I. Liquid Crystalline Solutions of Poly(Terephthalic Hydrazide) in Aqueous Organic Bases [a]

BASE	APPROXIMATE SOLUTION RANGE	
	WEIGHT PERCENT BASE	WEIGHT PERCENT POLYMER
TETRAMETHYLAMMONIUM HYDROXIDE	3 - 30	6 - 27
TETRAETHYL " "	7.5 - 25	7.4 - 26
DIETHYLAMINE	10 - 15	7.5 - 9
PIPERIDINE	10	7.5 - 9

(a) DETERMINED ON O-T AT 27°C. WHICH HAD η_{inh} OF 3.2-4.5 IN 5% AQUEOUS DIETHYLAMINE.

series of quaternary ammonium bases, sufficient
solubility for forming liquid crystalline solutions
was obtained only with tetramethylammonium hydroxide and tetraethylammonium hydroxide (Table I).

Recognition and Characterization of Liquid
Crystalline Solutions. The same visual and physical
characterization tests as used for aromatic polyamides (11) apply to recognition of liquid crystallinity in the polyhydrazide solutions. The
solutions appear somewhat turbid, often exhibit an
opalescent effect upon being stirred, depolarize
plane-polarized light, are oriented by a strong
magnetic field, and show a sharp decrease in bulk
solution viscosity above the critical concentration
point.

Critical Concentration and Saturation Regions
for Poly(terephtalic hydrazide). A typical critical concentration curve for O-T polymer is shown
in Fig. 4 for 10.4% $(CH_3)_4NOH$ and polymer with η_{inh}
of 4.4. Dilute isotropic solutions and aniso-
tropic solutions have very low bulk viscosities.
In regions above the critical concentration point
where both anisotropic and isotropic phases exist
together, phase separations can be produced upon
long standing or by centrifuging. The anisotropic
phase is the heavier, being higher in polymer concentration and containing a higher molecular weight
fraction.

Fig. 5 shows a plot of O-T polymer concentration versus tetramethylammonium hydroxide concentration. The lower line is a series of critical
concentration points which rises somewhat initially
with increasing base concentration and then becomes
nearly constant. The amount of polymer which can
be dissolved increases with increasing base concentration. In the saturation zone there is a
region wherein uniform solutions can be formed for

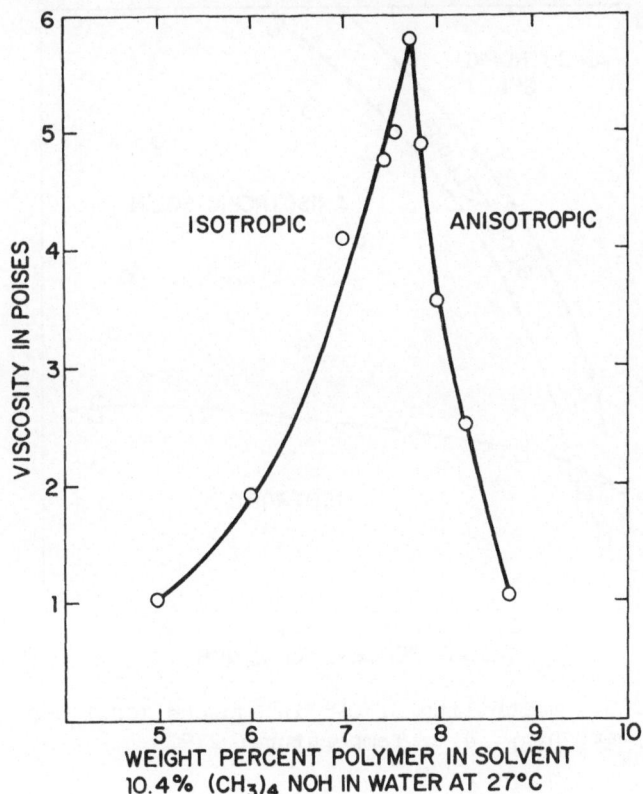

Fig. 4. Critical Concentration Curve for Poly(Terephthalic Hydrazide)(η_{inh} 4.4) in Aqueous Tetramethylammonium Hydroxide

a short time, but which are unstable and gel from a few minutes to several hours after preparation. There is a similar region in the isotropic zone at low base concentrations.

The range of solubility for O-T in tetraethylammonium hydroxide was much more restricted (Fig. 6). The slope of the saturation line is lower and the lowest critical concentration point is at a higher base concentration. The latter effect is due in part at least to the higher molecular weight of the

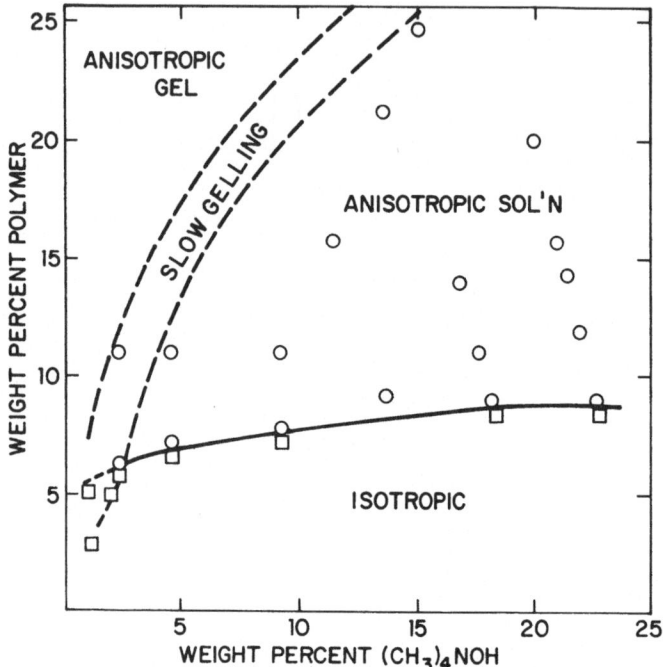

Fig. 5. Solubility of O-T in Aqueous $(CH_3)_4NOH$
Polymer: η_{inh}: 4.4, Temperature: 27°C

base. This point should occur at higher base concentrations for tetrapropylammonium hydroxide and
tetrabutylammonium hydroxide.

As with solutions of extended-chain polyamides
(12), the critical concentration is affected by
polymer molecular weight, being higher as molecular
weight decreases (Fig. 7). At low inherent viscosities solubility may be insufficient for the
attainment of a liquid crystalline state. On
the other hand, solvent-polymer interaction may be
so strong as to prevent the development of liquid
crystalline order.

Some samples of O-T, especially those with
viscosities below ~1.2, have been found to have low
or incomplete solubility. This effect may be due

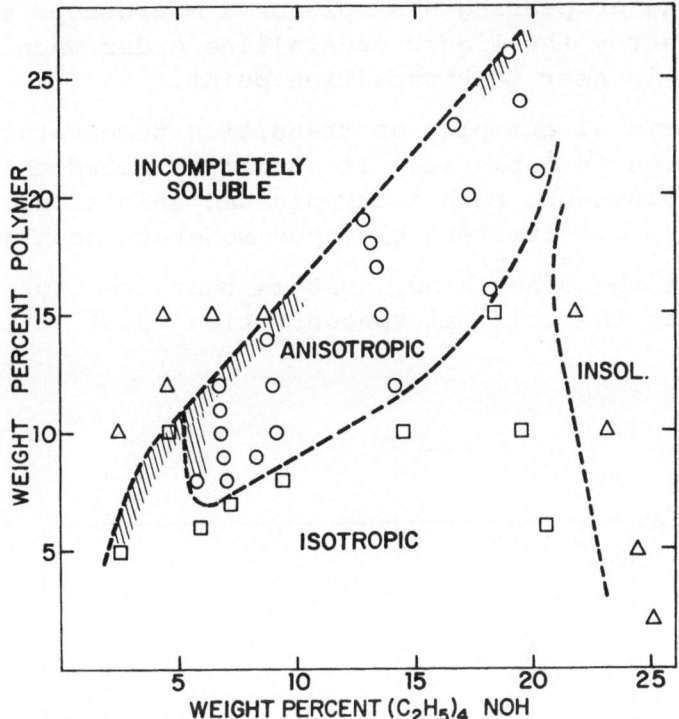

Fig. 6. Solubility of O-T in Aqueous $(C_2H_5)_4NOH$
Polymer η_{inh}:3.2, Temperature: 27°

to differences in crystallinity in the samples
associated with the method of preparation and isolation. Dobinson and Pelezo (13) mention a similar
difficulty with dissolution of some isolated
O-2/O-T copolymers in dimethylacetamide-LiCl.

Temperature Effect on Critical Concentration
Point. Nematic ↔ isotropic transitions were
much more readily effected by temperature changes
than had previously been evident for aromatic polyamide systems. Therefore, in determining critical
concentration points good temperature control is
important. A small temperature rise produced by

stirring or placing a sample on a microscope slide can destroy the liquid crystalline order when the system is near the transition point.

Several examples of transition temperatures are given in Table II. It should be noted that many solutions, both isotropic and anisotropic, may form gel irreversibly upon moderate heating.

On the other hand, systems that are isotropic and near the critical concentration point at room

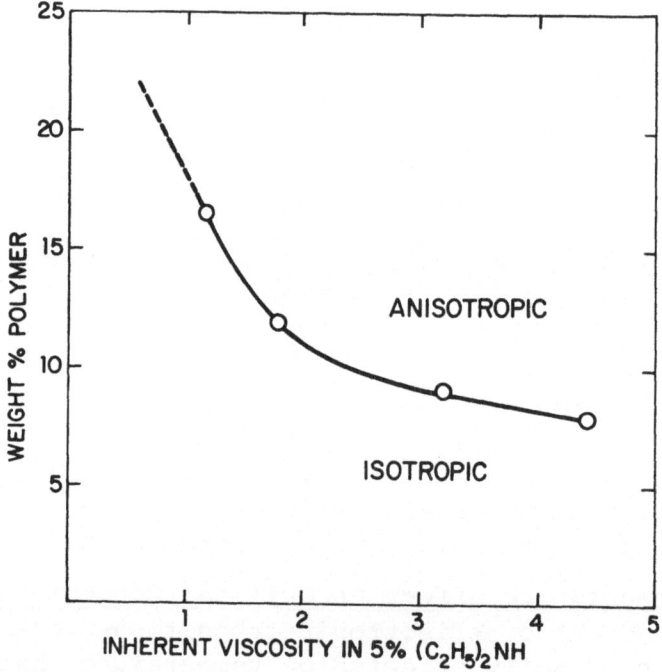

Fig. 7. Effect of Inherent Viscosity of Poly(Terephthalic Hydrazide) on the Critical Concentration in 25.5% Aqueous $(CH_3)_4NOH$

TABLE II. Nematic-Isotropic Transition Temperatures for Some Poly(Terephthalic Hydrazide) Solutions

| | POLYMER | | NEMATIC-ISOTROPIC |
SOLVENT	η_{inh}	% SOLIDS	TEMPERATURE, °C
10% $(C_2H_5)_4NOH$	3.2	9.1	34 - 36
10%	4.6	11.6	50 - 55
16.7% $(CH_3)_4NOH$	4.6	16.5	51 - 55
20%	4.6	19.1	52 - 54
20%	4.7	25.5	76 - 82

temperature (~25°C.) usually become optically anisotropic upon cooling to 0-10°.

Other Extended Chain Polyhydrazides. The polyhydrazide system based on terephthaloyl, chloro-terephthaloyl, 2,5-pyridinedioyl and oxalyl units was examined by the preparation of polymers and determination of their ability to form liquid crystalline solutions in aqueous tetramethylammonium hydroxide. The acyl groups are coded T, ClT, 2,5Pyr, and 2. A dihydrazide monomer unit is given by a code such as O-T-O. Thus, the mode of combination in the synthesis from hydrazides and diacid chlorides can be shown by the codes.

The polymers, inherent viscosities, and some solution tests are shown in Table III. O-2 and O-2,5Pyr polymers and copolymers with high proportions of these units were incompletely soluble in dimethyl sulfoxide-LiCl even in dilute solution. Therefore, aqueous diethylamine (5-20%) or 100% sulfuric acid was used for η_{inh} determinations.

TABLE III. Polyhydrazides and Their Solutions in Tetramethylammonium Hydroxide

POLYMER COMPOSITION [a]	η_{inh} [b]	SOLUTION IN AQUEOUS $(CH_3)_4NOH$		
		WEIGHT % BASE IN SOLVENT	WEIGHT % POLYMER IN SOLUTION	SOLUTION STATE
ONE AND TWO COMPONENTS				
(O-T-O)-T	4.72	15	10	ANISOTROPIC
O-T-O-(T/ClT)(90/10)	0.88	25	20	"
(80/20)	0.39	25	20	"
(50/50)	0.43	25	20	"
(O-ClT-O/O-T-O)-T (60/40)	0.92	25	20	"
(70/30)	1.43	25	20	"
(O-T-O)-ClT	1.16	25	20	"
(O-ClT-O)-T	1.07	25	25	"
O-ClT-O-(T/ClT)(50/50)	1.21	15	35 / ~40	ISOTROPIC / ANISOTROPIC
(O-ClT-O)-ClT	0.77	25	28	"
O-T-O-(T/2,5Pyr)(90/10)	0.17	25	20	"
(80/20)	0.64	25	20	"
(50/50)	0.43	25	20	"
(O-T-O)-2,5Pyr	0.70 [d]	25	10	"
(O-T-O/O-2,5Pyr-O)-T (70/30)	0.46	25	20	"
(60/40)	0.50	25	20	"
(O-2,5Pyr-O)-T	1.74 [c]	25	18	"
	3.95 [c]	17	19	"

TABLE III (Cont'd.)

POLYMER COMPOSITION [a]	η_{inh} [b]	SOLUTION IN AQUEOUS $(CH_3)_4NOH$		
		WEIGHT % BASE IN SOLVENT	WEIGHT % POLYMER IN SOLUTION	SOLUTION STATE
(O-T-O/O-2,5Pyr-O)-2,5Pyr (60/40)	0.53	25	30	ANISOTROPIC
(50/50)	1.09	25	10-20	ISOTROPIC-INSOL.
(O-2,5Pyr-O)-2,5Pyr	0.86[d] 0.84[c]	15	20	ISOTROPIC AND GEL ISOTROPIC AT 60°
(O-ClT-O)-2,5Pyr	1.43 1.98	25	29	ANISOTROPIC ISOTROPIC AT 50°
(O-ClT-O/O-2,5Pyr-O)-2,5Pyr (50/50)	1.04	25	20	ANISOTROPIC
(O-2-O)-2	0.31	25	33	ISOTROPIC
(O-2-O/O-T-O)-2 (50/50)	0.49[d]	15	20	"
(12.5/87.5)	0.38[d]	15	20	"
(O-2-O)-T	0.24[d]	25	33	**ANISOTROPIC**
(O-2-O/O-ClT-O)-2 (50/50)	0.33	25	33	ISOTROPIC AT 0°
(O-2-O)-ClT	2.06	25	15	ANISOTROPIC
(O-2-O)-2,5Pyr	0.98[d] 0.56[c]	25 15	25 20	ISOTROPIC AND GEL "
THREE COMPONENTS				
O-2-O-(T/ClT) (90/10)	1.05[d] 0.92[c]	25	20	ANISOTROPIC
(80/20)	1.10[d]	25	20	"
(70/30)	0.35[d] 0.70[c]	25	30	" at 20°

TABLE III (Cont'd.)

POLYMER COMPOSITION [a]	η_{inh} [b]	WEIGHT % BASE IN SOLVENT	WEIGHT % POLYMER IN SOLUTION	SOLUTION STATE
O-2-O-(T/ClT) (60/40)	0.49[c] 0.57[d]	25	20 27	ISOTROPIC GEL
(50/50)	0.38[c]	25	29	ISOTROPIC
O-2-O-(T/2,5Pyr) (90/10)	0.63[d] 1.03[c]	25	33	ANISOTROPIC AT 20-25°
(75/25)	0.59[d] 0.72[c]	15	20	ISOTROPIC
(50/50)	0.48[c]	25	27	"
(O-2-O/O-2,5Pyr-O)-T (70/30)	1.18	25	20	ANISOTROPIC
(O-2-O/O-T-O/O-2,5Pyr-O)-T (50/20/30)	0.69	25	33	"
FOUR COMPONENTS				
(O-2-O/O-T-O/ClT/2,5Pyr) (25/25/25/25)	0.82	25	28	"

(a) 2, T, ClT, AND 2,5Pyr REPRESENT OXALYL, TEREPHTHALOYL, CHLOROTEREPHTHALOYL AND 2,5-PYRIDINEDIOYL UNITS INTRODUCED AS DIACID CHLORIDES; O-2-O, O-T-O, O-ClT-O AND O-2,5Pyr-O ARE CORRESPONDING DIHYDRAZIDE UNITS, INTRODUCED AS THE DIHYDRAZIDES.
(b) DETERMINED IN DIMETHYLSULFOXIDE-5% LiCl UNLESS OTHERWISE NOTED.
(c) 5% AQUEOUS DIETHYLAMINE.
(d) 100% H_2SO_4 AT 25°.

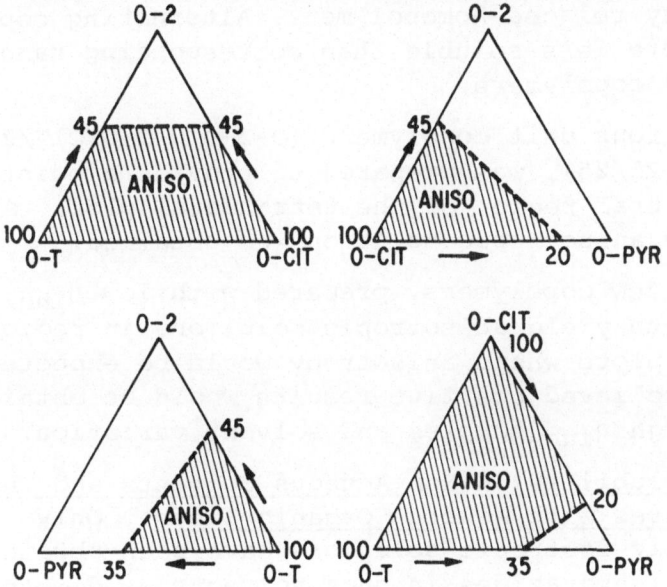

Fig. 8.   Regions of Observed Anisotropic Solution Formation for Polyhydrazides in Aqueous $(CH_3)_4NOH$

Fig. 8 provides plots of solubility tests on polymers with varying molecular weights. These plots form the faces of an equilateral tetrahedron and indicate the region in which liquid crystalline solutions have been obtained.

The failure of poly(oxalic hydrazide) and polymers with high proportions of oxalyl units to yield liquid crystalline solutions is partly due to insufficient solubility in the solvents tested. Inadequate molecular weight may be a factor also.

At comparable molecular weights the solubility of the homopolymers in aqueous bases is judged to be O-ClT > O-T > O-2,5Pyr > O-2. Copolymers, in general, were more readily soluble than the least soluble, related homopolymer and often more soluble

than any related homopolymer. Alternating copolymers were less soluble than corresponding randomly
ordered copolymers.

A four unit copolymer, (O-2-O/O-T-O/ClT/2,5Pyr)
(25/25/25/25), was prepared to provide a point in
the central region of the tetrahedral plot. An optically anisotropic solution was obtained.

A few copolymers, prepared with low η_{inh},
failed to yield anisotropic solutions in regions
of the plots where anisotropy would be expected.
It is believed positive results would be obtained
from high η_{inh} samples and solvent variation.

Solubility in Non-Aqueous Solvents and Their
Admixtures with Aqueous Organic Bases. Only
optically isotropic solutions have been obtained
with polyhydrazides in such solvents as dimethyl
sulfoxide-LiCl or dimethylacetamide-LiCl.

Although some sulfoxides and amides in limited
amounts can be added to and are compatible with
polyhydrazide solutions in aqueous organic bases,
liquid crystalline order is often destroyed by minor
amounts of these additives. For example, an optically anisotropic, 19.7% solution of O-T (η_{inh} 4.4)
in 25% tetramethylammonium hydroxide diluted with
~5% dimethyl sulfoxide so that the polymer content
was 18.8% (by weight) was anisotropic. Upon doub-
ling the amount of dimethyl sulfoxide so that the
polymer concentration was at 17.8%, the solution
became isotropic. The polymer concentration was
far above the critical point range of ~10% (Fig. 5)
for a water-based system. The yellow coloration
was not discharged by the addition but the solution
viscosity was much increased. Presumably the added
polar solvent disrupts the association between organic base and polymer sufficiently to produce a
more flexible chain and decrease the persistence
length.

Similar effects are obtained with other water-miscible solvents such as methanol, tetrahydrofuran, and dioxane. In these cases polymer precipitation may be encountered at moderate amounts of additive.

No optically anisotropic solutions have been formed from organic bases in organic solvents without water present.

Extended Chain Polymers Without Ring Components. Up to this point, with one exception, i.e., poly(oxalic hydrazide), only ring-containing polymers have been considered. The poly(oxalic hydrazide) available, having an η_{inh} of 0.31, gave isotropic solutions in the systems tried. This polymer in higher molecular weight should yield anisotropic solutions if adequate solubility is retained.

Another extended-chain polyhydrazide which contains no rings was prepared from oxalic di-hydrazide and trans,trans-muconyl dichloride (last formula of Table IV). The polymer had an η_{inh} value of 0.33 and dissolved in tetramethylammonium hydroxide to form only isotropic solutions. With a higher molecular weight sample, optically aniso-tropic solutions should be attained. Other possible components for this purpose are fumaroyl and acetylenedioyl (Table IV).

Polyhydrazides with Meta-Oriented Ring Units. A wide variety of polyhydrazides are soluble in aqueous solutions of selected organic bases. However, the concentrated solutions are isotropic unless the proportion of meta-linked or flexible units is relatively small (i.e., less than about 20 mole percent). Examples of polymers which dissolve to form isotropic solutions only are shown in Table V.

TABLE IV.   Some Non-Ring-Containing Polyhydrazides with Extended Chain Potential

TABLE V.   Some Isotropic Solutions of Polyhydrazides

POLYMER		SOLUTION	
COMPOSITION	η_{inh} [a]	$(CH_3)_4NOH$	WEIGHT PERCENT POLYMER
O-I	0.76	10%	20
		15%	25
O-I-O-tBuI	0.51 [b]	25%	33
O-I-O-T	0.40	10%	20

(a) DIMETHYLSULFOXIDE-5% LiCl
(b) m-CRESOL

Isotropic solutions are obtained likewise from polyhydrazides containing large proportions of linear polymethylene units, such as adipoyl or sebacoyl units.

Copolyhydrazide-amides. Polyhydrazides with high proportions of amide links are poorly soluble in aqueous organic bases or at best yield isotropic solutions. As an example, O-T polyhydrazide with 5 and 10 mole percent amide links derived from p-phenylenediamine formed optically anisotropic solutions in 10% and 25% tetramethylammonium hydroxide. At 20% modification, or above, only isotropic solutions or incomplete solubility was attained (Tables VI and VII).

TABLE VI. Solubility of Poly(Terephthalic Hydrazide-co-p-Phenylene Terephthalamides) in 10% Aqueous $(CH_3)_4NOH$

COPOLYMER		SOLUBILITY TEST				
HYDRAZIDE/	η_{inh}	WEIGHT PERCENT POLYMER				
AMIDE RATIO	dl/g	5	10	15	20	30
95/5	2.41^a		ISO	ANISO	ANISO	GEL
90/10	2.77^a	ISO	ANISO	ANISO		
80/20	0.40^b	PARTLY INSOL.	GEL			
70/30	0.37^b	PARTLY INSOL.				

a - 10% $(CH_3)_4NOH$
b - H_2SO_4; DEGRADED.

TABLE VII.   Solubility of Poly(Terephthalic Hydrazide-co-p-Phenylene Terephthalamides) in 25% Aqueous (CH$_3$)$_4$NOH

COPOLYMER		SOLUBILITY TEST				
HYDRAZIDE/	ηinh		WEIGHT PERCENT POLYMER			
AMIDE RATIO	dl/g	5	10	15	20	30
95/5	2.44[a]		ISO	ANISO	ANISO	
90/10	2.77[a]		ISO	ANISO	ANISO	ANISO
80/20	0.40[b]	PARTLY INSOL.	GEL	GEL	PARTLY INSOL.	
70/30	0.37[b]	PARTLY INSOL.	GEL			

a - 10% (CH$_3$)$_4$NOH
b - H$_2$SO$_4$; DEGRADED

N-Substituted Polyhydrazides.   Several poly-hydrazides were prepared in which each hydrazide link had one N-methyl substituent (Table VIII). Only isotropic solutions were obtained in the aqueous organic bases.   The nitrogen substituent presumably provides a more flexible type of amide link and changes the nature of the solvent association (see next section).

Coloration in Polyhydrazide-Base Solutions. The wholly aromatic, para-linked polyhydrazides exhibit a deep yellow coloration in aqueous bases. This coloration is not shown in polymers wherein the aromatic rings are all meta-linked, or poly-hydrazides derived from monomethylhydrazine (Table VIII).   Intermediate color formation is shown by such polymers as alternating O-I-O-T and

TABLE VIII.   Isotropic Solutions of Polymers Derived From Terephthalic Bis(Monomethylhydrazide)

COMONOMER UNIT	POLYMER η_{inh}[a]	SOLUTION	
		$(CH_3)_4NOH$ IN WATER	WEIGHT PERCENT POLYMER
TEREPHTHALOYL	0.38	25	29
CHLOROTEREPHTHALOYL	0.37	15	25
2,5-PYRIDINEDIOYL	0.25	15	25

(a) IN DIMETHYLSULFOXIDE-5% LiCl.

O-2-O-T. The color appears to be the result of keto-enol tautomerism of the carbonyl groups and the consequent extension of conjugation in the enol form, for the color in a basic solution is discharged reversibly by precipitation of the polymer in aqueous acid.

Some polymers may have a deep yellow color as prepared. This is probably due to the presence of residual inorganic or organic base not removed in the polymer work-up. Dissolution followed by precipitation in slightly acid media will remove the color.

Fig. 9 shows the visible-ultraviolet light absorption spectra obtained for dilute solutions of O-T, O-I-O-T and (MeO-T-MeO)-T in 10% aqueous $(CH_3)_4NOH$. The yellow color decreases in the order given above.

Raman spectra were obtained for a 9.1% solution of O-T in 10% tetraethylammonium hydroxide, solid

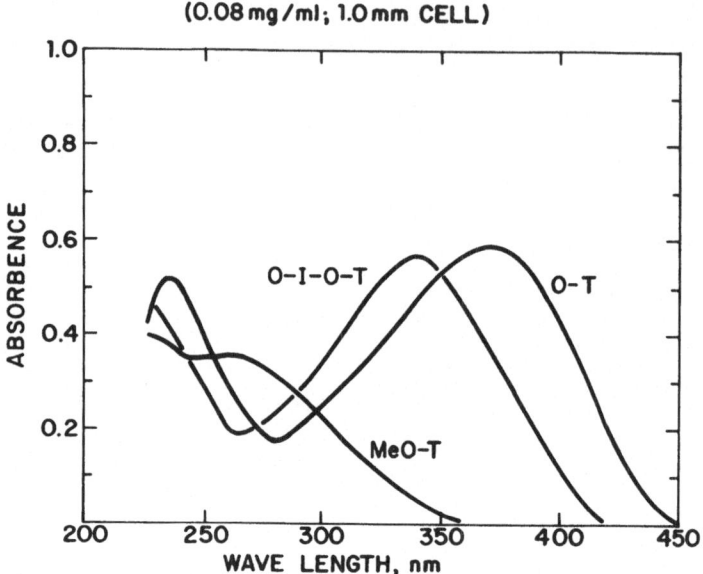

Fig. 9. Ultraviolet-Visible Light Absorption Spectra for Polyhydrazides in Aqueous 10% Tetramethylammonium Hydroxide Solution

O-T and dibenzoylhydrazine using the 4880Å argon line as excitation source.

Comparison of the spectrum of solid O-T with that of the solution showed the disappearance of the amide carbonyl at 1685 cm.$^{-1}$ and the appearance of new intense bands at 1500 and 1270 cm.$^{-1}$. Dilution of the solution with solvent had no significant effect on the spectrum. Comparison of the spectrum with that of a saturated solution of sodium benzoate in 10% aqueous tetraethylammonium hydroxide showed the absence of carboxyl groups in the O-T solution. Therefore, hydrolysis was absent. The spectrum of dibenzoylhydrazine in 10% base was similar to those of O-T solutions. The spectrum of O-T in 12.5% aqueous diethylamine was identical to that in 10% aqueous tetraethylammonium hydroxide.

The intensity of the Raman bands in the spectra of O-T and dibenzoylhydrazine solutions indicated that a higher degree of conjugation, relative to the starting materials, had resulted. In view of the spectral evidence for loss of carbonyl structure and the presence of extended conjugation, the presence of the following structural unit in the polymer chain in solution is suggested (Fig. 10).

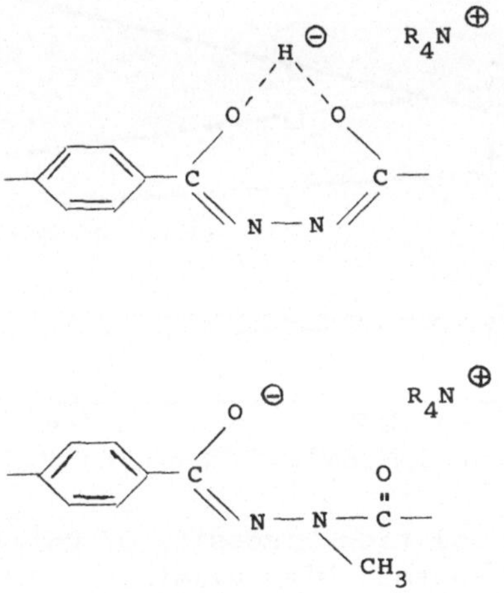

Fig. 10. Possible Chain Unit Structures of Polyhydrazides Dissolved in Aqueous Organic Base

N-Monosubstituted hydrazides, which do not exhibit visible color, are presumably simple enol salts. Polymers of this type yield only isotropic solutions (Table V).

Dilute Solution Behavior. Dilute solution viscosities were determined in 5% aqueous diethyl-amine at varying concentrations for several samples of O-T (Fig. 11). Straight line plots were obtained, which indicate the absence of a polyelectrolyte effect which might be present because of the

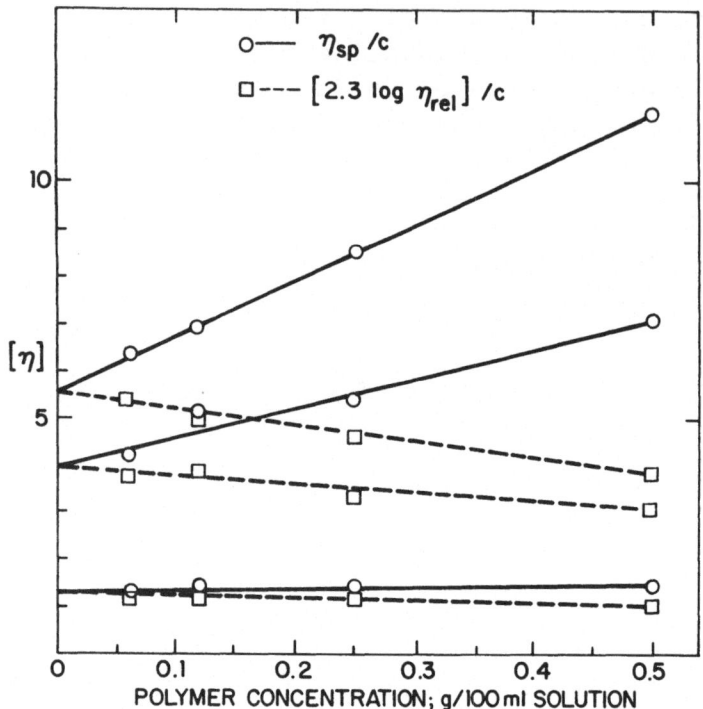

Fig. 11.   Dilute Solution Viscosity of Poly(Terephthalic Hydrazide) in 5% Aqueous Diethylamine at 25°C

existence of the polymer in an ionized enol form (preceding section).

The Huggins' constants for these solutions were calculated to be 0.37 and 0.38 for the upper two plots, but only 0.14 for the lower plot.   The latter is abnormally low.

Dilute solutions in aqueous diethylamine up to at least 20% likewise gave linear plots at slightly lower viscosities.   Similar results were obtained for dilute solutions in aqueous tetramethylammonium hydroxide.

Polymer Stability in Base Solutions. There
was little evidence of polyhydrazide degradation
in normal handling of dilute or concentrated
solutions. Upon storage of concentrated solutions
for several months at room temperature, marked
degradation has been noted. Solutions of poly(terephthalic hydrazide) (9.1% in 10% tetraethylammonium
hydroxide) showed a slight loss in inherent viscosity when wet spun into fibers in a process
employing acidic coagulation baths (η_{inh} from 3.8
to 3.5 and 4.5 to 4.2).

Fiber and Film Preparation

Strong fibers and films were prepared from O-T
solutions by coagulation in aqueous phosphoric or
acetic acid baths. For example, fibers with denier/
tenacity/elongation/initial modulus of 2.04/10.5
gpd/13.7%/220gpd were obtained from a 9.1% solu-
tion of O-T (η_{inh}5.1) in 10% $(C_2H_5)_4NOH$ spun into a
1/21.8 phosphoric acid/water bath at 19°C. After
passing the dry fibers over 34 cm. of a hot metal
surface at 200°C. at 6 m./min. with 10% extension,
the denier/T/E/Mi were 1.96/12.7 gpd/7.4%/363 gpd.
The fibers had orientation angles as-spun of 28°
and heat-treated, 10°, as determined by x-ray
diffraction. The density of similar fibers was
1.458 g./cc.

EXPERIMENTAL

Intermediates

Oxalic dihydrazide (m.p. 241-243.5°) was pur-
chased from Eastman Organic Chemicals and used as
received. Other dihydrazides were prepared from
dimethyl esters and hydrazine as illustrated below.

Methyl chloroterephthalate melted at 56–57° and
the dihydrazide at 217–219°.

Methyl Ester of 2,5-Pyridine Dicarboxylic Acid:
2,5-Pyridine diacid chloride (50 g.) was added slowly to 600 ml. of methanol with stirring. The solution was warmed on a steam bath for 4 hours and
then allowed to stand overnight without heating. A
white crystalline product formed. The alcohol was
reduced to half volume by evaporation and the
product collected and dried; m.p. 161–162°.

2,5-Pyridine Dicarboxylic Hydrazide: A mixture
of 100 ml. of benzene, 15 g. of hydrazine hydrate,
and 28 g. of 2,5-pyridine dicarboxylic acid methyl
ester was heated at reflux on a steam bath for
24 hrs. The mixture was cooled and diluted with
200 ml. of methanol and the finely crystalline
product collected. This was dissolved in 1500 ml.
of water at the boil, and decolorized with charcoal.
About half of the water was removed by evaporation
and the product allowed to crystallize. The yield
was 22.6 g.; m.p. 279–280. Anal.: Calcd. for
$C_7H_9O_2N_5$: C, 42.9; H, 4.62; N, 35.7. Found:
C, 42.9, H, 4.88; N, 35.2.

Terephthalic Bis(monomethylhydrazide): A
mixture of 600 ml. of benzene, 200 g. of dimethyl
terephthalate, and 92.1 g. of monomethylhydrazine
was heated at reflux on a steam bath for 64 hrs.
Product precipitated from solution. Methanol
(500 ml.) was added to the cooled mixture. The
crystalline product was collected by filtration,
washed with dilute aqueous sodium carbonate, and
then water. The yield of hydrazide was 50 g.,
melting at 240–241°.

Anal.: Calcd. for: $C_{10}H_{14}O_2N_4$: C, 54.0;
H, 6.30; N, 25.2. Found: C, 54.0; H, 6.37; N,
25.4.

An nmr spectrum confirmed a symmetrical structure with the methyl substituents on the terminal nitrogen atoms.

Polymer Preparation

A suspension of terephthalic dihydrazide (9.71 g.; 0.05 mole) in 250 ml. of dimethylacetamide containing 8.40 g. lithium chloride was cooled to 0°C. in an ice-water bath. Chloroterephthaloyl chloride (11.88 g.; 0.05 mole) was added over a period of 70 minutes, while the mixture was stirred and protected from moisture with a flow of dry nitrogen. The solution became very viscous and was stirred an additional 30 minutes. The dope was neutralized with lithium carbonate (3.69 g.; 0.05 mole). An isolated, washed and dried sample of polymer had an η_{inh} of 3.45 (dimethyl sulfoxide - 5% LiCl). The yield of dry polymer was 98%.

Inherent Viscosity Determination

Inherent viscosity $[\eta_{inh} = \ln(\eta_{rel})/c]$ was determined at 30°C. and a polymer concentration (c) of 0.5 g./100 ml. of solution. The traditional solvent was dimethyl sulfoxide with 5% by weight of lithium chloride. However, not all polyhydrazides will dissolve in this solvent. More generally useful viscosity solvents are the aqueous organic bases. For this purpose 5% by weight of diethylamine in water was used. Alternatively, 100% sulfuric acid at 25°C. can be used for polymers with poor solubility.

ACKNOWLEDGMENT

Thanks are due to W. F. Dryden, Jr. for excellent technical assistance.

BIBLIOGRAPHY

1. A. H. Frazer and F. T. Wallenberger, J. Polymer
    Sci., A2, 1137, 1147, 1171 (1964).

2. A. H. Frazer, W. Sweeny, and F. T. Wallenberger
    J. Polymer Sci., A2, 1157 (1964).

3. J. Preston and W. B. Black, J. Polymer Sci.,
    B4, 267 (1966); C19, 17 (1967).

4. J. Preston and R. W. Smith, J. Polymer Sci.,
    B4 1033 (1966).

5. W. B. Black and J. Preston, "High Modulus
    Wholly Aromatic Fibers", Marcel Dekker, Inc.,
    New York, N.Y., 1973.

6. B. M. Culbertson and R. Murphy, J. Polymer
    Sci., B4, 249 (1966); B5, 807 (1967).

7. A. H. Frazer, U.S. Patent 3,642,707 (2/15/72)
    assigned to the Du Pont Company.

8. A. Ciferri, Polymer Eng. Sci., 15 (3), 191
    (1975).

9. G. Allegra, Polymer Eng. Sci., 15 (3) 207
    (1975).

10. J. D. Hartzler, U.S. Patent 3,966,656 (6/29/76),
    assigned to the Du Pont Company.

11. P. W. Morgan, Polymer Preprints, 17 (1), 47
    (1976); paper presented at the meeting of
    the American Chemical Society, New York, N.Y.

12. S. L. Kwolek, P. W. Morgan, J. R. Schaefgen,
    and L. W. Gulrich, Polymer Preprints, 17 (1),
    53 (1976); paper presented at the meeting of
    the American Chemical Society, New York, N.Y.

13. F. Dobinson and C. A. Pelezo, U.S. Patent
    3,748,298 (7/24/73), assigned to the Monsanto
    Company.

DISCUSSION

F. H. WINSLOW - BELL LABORATORIES. What effects do transition metal salts have on these polymers, e.g., will they chelate copper ions? P. W. MORGAN - DU PONT (RETIRED). We haven't done that, so I can't answer the question with data. I think that the polyhydrazides might chelate such metals and hold them well, although some metals and the chelates might not stay in solution.

J. R. SCHAEFGEN - DU PONT. I noticed in the solubility curves at the isotropic-anisotropic boundary with the aqueous tetramethylammonium hydroxide and the O-T polyhydrazide that the amount of base only increased slightly as the amount of polymer in solution increased. One would think that if the postulated ring structure formed, to keep it formed at every hydrazide group one would have to add correspondingly more base as polymer concentration increased. This is the case with tetraethylammonium hydroxide solution, the curve increasing steeply in base as polymer solubility increased. Evidently about one base molecule per hydrazide link is required with tetramethylammonium hydroxide at the initial part of the curve but little more to confer the required rigidity for anisotropic solution formation as polymer concentration increases, while in the case of tetraethylammonium hydroxide you require about one mole of base per hydrazide unit over a wide range of polymer concentration. Can you account for this difference in behavior of these two solvents? P. W. MORGAN. There are, as you say, major differences in the zones for liquid crystal formation for the O-T in the two solvents. First, the inherent viscosity of the polymer used in the two cases is different but not enough to account for the point in question. The line for a 1 to 1 mole ratio of tetramethylammonium hydroxide to hydrazide units in O-T is close to but somewhat below the solution saturation line (Figure 5), and far above the line of critical concentration points. Therefore, there is a large excess of base, relative to the 1/1 ratio, at the critical concentration points except at the lower base concentrations. In the case of tetraethylammonium hydroxide (Figure 6), the line for a 1/1 mole ratio parallels the saturation line but is appreciably below it and closer to the critical concentration line. One might argue that this solvent is more

effective or efficient in dissolving the polymer in a
narrow range. The strong polymer-solvent association may
prevent liquid crystal formation until the system is
diluted with polymer. This gives rise to the higher slope
of the critical concentration line. The size of the
solvent cations, their hydrocarbon character, and the
dissociation of the base in water may also be factors.

FRANK DOBINSON - MONSANTO. Both Monsanto and
Dr. Culbertson at Ashland Oil Co. have worked on a polymer
from p-aminobenzhydrazide and terephthaloyl chloride which
is a homopolymer with copolymer attributes due to the
asymmetry of one of the monomers. The polymer is 50% amide
and 50% hydrazide, obviously, and according to your graph I
assume that it would not form anisotropic solutions under
the conditions that you have looked at. Is that correct?
P. W. MORGAN. Yes, we have looked at such a polymer and it
does not dissolve well enough; in fact, it dissolves very
little in any base system we have tried. The two slides on
solubility of random polyamide-hydrazides showed that, with
as little as 20% amide units, solubility in the bases was
poor.

C. OVERBERGER - U. MICHIGAN. As I understand it,
Paul, you suggest that you have a strong enough base to
make an anion of the enol of the amide group and also that
the gegenion has some influence on the structure.
Presumably, potassium hydroxide would work as well as
the tetramethylammonium hydroxide if the solubility factors
were right.   P. W. MORGAN. That is correct. These
polyhydrazides are somewhat soluble in bases such as
potassium hydroxide, perhaps 2 or 3% at the most. I refer
you also to British Patent Specification 1,209,207 (Oct.
21, 1970). You may remember that Van Krevelen and others
of Akzo Research Laboratories worked with poly(terephthaloyl-
oxalamidrazones). These polymers dissolved quite well in
inorganic bases, but, as far as I know, there has been no
report of liquid crystal formation.

C. OVERBERGER. The curve of weight percent polymer vs.
weight percent tetraalkylammonium hydroxide was at 27°C.
If the plot were made at 50°C, would the anisotropic portion
be much larger?   P. W. MORGAN. At higher temperatures,
e.g., 50°C, the solution must be more concentrated to reach
the anisotropic boundary. The boundary line will be raised

but the anisotropic region will be smaller because you do
not increase overall solubility in the same way; in fact
you may decrease it and observe more gel. Thus, if you go
too high in temperature, you may wipe out the anisotropic
region entirely.

F. BAILEY - UNION CARBIDE. What happens if you use
a "crown ether" base of potassium as a potential solvent?
P. W. MORGAN. I don't know. We haven't done that; it
would be an interesting area to explore.

I. FALKEHAG - WESTVACO RESEARCH. Would you elaborate
on the nature of the complex formation reaction between
base and hydrazide.   P. W. MORGAN. The structure shown
was a cyclic one resulting overall from enolization of both
amide groups in the hydrazide unit and interaction of this
structure with an organic amine hydroxide. As depicted,
one hydrogen is shared or associated with the two
negatively charged oxygens. This loosely formed cyclic
structure has a negative charge balanced by a positive
ammonium gegenion. The other H^+ is associated with an OH^-
of the medium, or is present as a water molecule. The
cyclic structure with one associated base molecule was
proposed because at one time in our work it appeared that
optimum solubility is obtained at that ratio. Today, I am
less convinced that the ratio is optimum under all
circumstances. I do not believe that a cyclic structure
is needed to obtain liquid crystallinity. One only needs
the trans configuration in the amide or hydrazide units,
via double-bond character or otherwise, for an extended
chain.

M. BANKIEWICZ - CHEMICAL ABSTRACTS SERVICE. My question concerns the time and mechanical shear effects on the liquid crystalline solutions. How long does it take for crystallinity to appear when you pass the solution through a tube or an orifice? Does it help the alignment of the crystals? This would be of interest in the production of these liquid crystal solutions.   P. W. MORGAN. Yes, shear does improve the alignment of the crystals. First, exceeding the critical concentration alone produces liquid crystallinity spontaneously; stirring may be used to speed solution but liquid crystallinity remains upon resting.

M. BANKIEWICZ. What is the order of magnitude of the effect, microseconds or minutes?   P. W. MORGAN. Some response is evident in extremely short times; for example, a minor disturbance of a sample in a polarizing microscope produces an instantaneous change in birefringence pattern. Alignment of liquid crystalline domains depends on the shear. Relaxation is fairly fast but full relaxation might take a minute or hours, depending on fluidity and other variables. Compared to low molecular weight, thermotropic liquid crystal systems, the polymeric solutions have a very slow response.

A. B. CONCIATORI - CELANESE CORP. I would like to ask about some of the properties you reported on the fibers. You said you spun from a liquid crystalline solution and obtained 10 g/d. Do you attribute that value to the structure of the liquid crystalline solution? If you spun normally from a non-liquid crystalline state how would the result compare with the above value?   P. W. MORGAN. If you spin a randomly organized solution such as from an isotropic solution in dimethyl sulfoxide-LiCl, then the tenacity would not be over 2 g/d either by dry or wet spinning techniques, i.e., just extrusion and windup. But if you draw in the bath or afterwards, and this is possible to some degree, especially by complicated processes such as staging, you can improve tenacity appreciably. Fibers from liquid crystalline solutions can come out highly oriented upon simple extrusion whereas orientation from isotropic systems is undetectable by X-ray diffraction.

A. B. CONCIATORI. Because of the structure of the fibers obtained by spinning from lyotropic solutions, can they be converted more effectively to polyoxadiazole than polyhydrazides originally prepared by A. H. Frazer of

Du Pont?   P. W. MORGAN. Conversion by dehydration to an
oxadiazole is a fairly difficult, slow process. I am not
aware of any increase in conversion rate due to a different
degree of order in these fibers. I do know it takes several
hours, and usually tenacity of the fibers goes down after
conversion.

F. H. WINSLOW. How does an electric field affect the
orientation of these systems? Is it useful in the spinning
operation? And are these systems of any value for display
devices?   P. W. MORGAN. The systems do orient in an
electric or a magnetic field but it is a relatively slow
process and it is difficult to apply to a spinneret through
which a solution is moving at a high speed. The systems do
not become oriented or relax rapidly compared to the rates
for low molecular weight liquid crystal systems.

F. H. WINSLOW. You don't observe any synergism with
the combination of shear and an electric field?
P. W. MORGAN. No, nothing that is really helpful in
spinning.

R. T. NATARAJAN - LORD CORP. My question is related
to the previous one, but asks the effect of a field before
and during the polymerization. Will you see any effect on
orientation by an electric or magnetic field applied during
polymerization?   P. W. MORGAN. The monomers should not
respond very well but as polymerization proceeds, there
might be an effect. We haven't done that. However, you
can't polymerize in a quaternary ammonium base solution and
the amide media which we have used for polymerization have
not yielded liquid crystalline solutions.

R. T. NATARAJAN. What is the role of phosphoric acid
in the spinning system?   P. W. MORGAN. It is added to
the bath in wet-spinning to neutralize the base quickly to
get coagulation. If you spin into water, you will go
through a slow process of getting a more and more dilute
base solution - but a polymer solvent nevertheless, so
the step from polymer solution to a coagulated state is
relatively slow. The result could be a very soft, swollen
fiber or perhaps no fiber at all.

R. T. NATARAJAN. Can you compare the properties of these materials with Kevlar® aramid fiber properties?
P. W. MORGAN. The "Kevlar" aramid fiber that is on the market has a tenacity somewhere in the 20 g/d range, i.e., 20-25 g/d; so the polyhydrazide fiber described has about half that value.

R. BAUGHMAN - ALLIED CHEMICAL. Can you detail again the changes in Raman spectra which are observed in going between the isotropic and the nematic phases?
P. W. MORGAN. In the solid state carbonyl bands at 1685 cm^{-1} are observed, while in 10-20% solution in base there are no carbonyl bands at all. At the same time major new bands appeared at 1500 and 1270 cm^{-1}, indicating conjugation.

R. BAUGHMAN. I meant the changes between the isotropic solution and the nematic phase solution. P. W. MORGAN. We have not observed any differences between isotropic and nematic solutions. I would not expect the spectra of isotropic and nematic solutions to differ except insofar as they are affected by orientation and concentration. The polymer chains in the isotropic solution are presumably fully extended and associated with the solvent in much the same way as in the nematic phase. The transition from isotropic to anisotropic state, which occurs at the critical concentration, but incompletely, does not require the chains to become more fully extended. One could look for changes in solvent association for isotropic solutions in base of varying concentration.

R. BAUGHMAN. The Raman spectroscopy of the phase changes should be quite interesting, especially since the colorless solid polymer forms yellow nematic phase solutions. By choosing the excitation wavelength in the absorption region of the solution, resonance enhanced spectra should be obtained. I would expect that these colored polymer solutions are likely to be fluorescent. Have you observed fluorescence? P. W. MORGAN. I didn't do this part of the work myself and don't know whether fluorescence was observed.

R. S. MOORE - EASTMAN KODAK. Is there evidence of block structures in the various copolyhydrazides which you have described?   P. W. MORGAN. Many of the two-component polymers were alternating copolymers. I have no evidence of blockiness in any of the others. Those polymers which remained in solution during preparation should have a random distribution of units even with differing reaction rates for the components.

PROPERTIES OF RIGID-CHAIN POLYMERS IN DILUTE AND CONCEN-

TRATED SOLUTIONS

Guy C. Berry

Department of Chemistry
Carnegie-Mellon University
Pittsburgh, PA   15213

SYNOPSIS

The salient features of investigations on several
polymers characterized by long, rigid sections in the chain
skeleton are discussed. Procedures to estimate the persistence length ρ, the optical anisotropy δ and certain
other molecular parameters are considered and the results
for several examples are given. The effect of intermolecular aggregation on estimates for these parameters is
important in some cases. Some rheological properties of
concentrated solutions of certain of the polymers are discussed. The behavior of both optically isotropic and
anisotropic solutions of rodlike polymers is included,
along with the properties of a chain that can be described
as a rigid coil molecule. Emphasis is on the dependence
of the steady state viscosity η_K on shear rate κ, the
recovery function R_K for recoverable strain γ_K^R after cessation of steady state flow ($R_K = \gamma_K^R/\kappa\eta_K$) and the creep
behavior.

INTRODUCTION

In recent years the properties of several polymers
characterized by long, rigid chain skeletons have been
investigated in our laboratory[1-12]   In this report, we
will summarize the current status of some of this work
dealing with the behavior of such polymers in dilute and
concentrated solutions. The structure of the polymers
we will discuss is given in Table I.

TABLE I

Structure of Polymers Discussed in Text

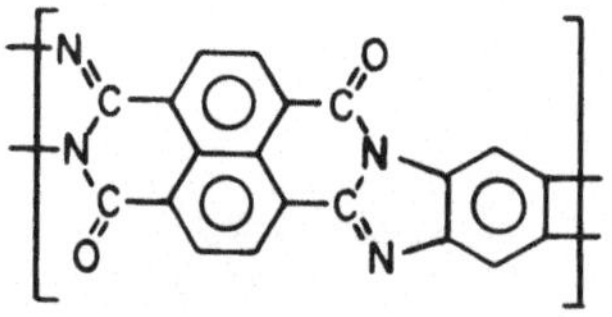

BBL

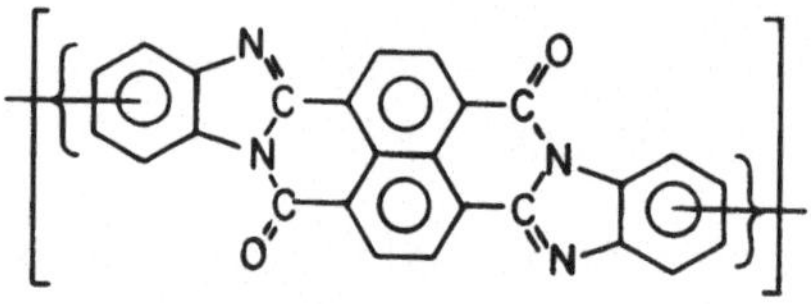

BBB

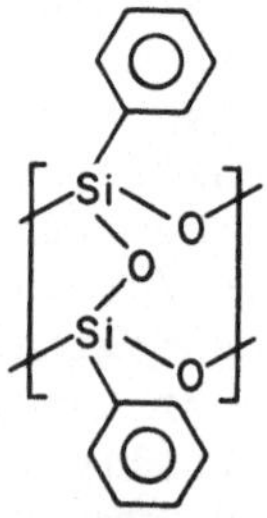

PPSQ

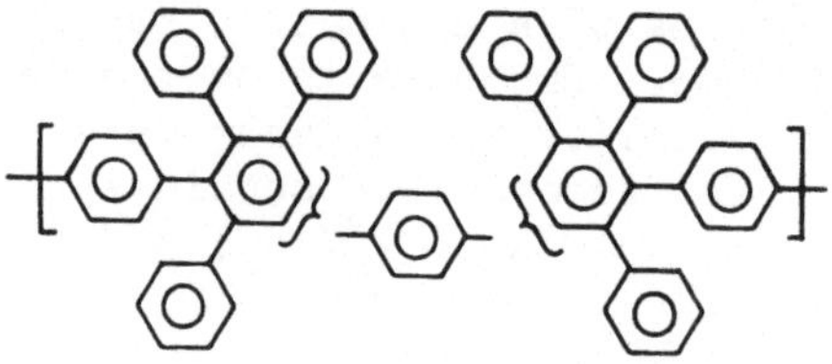

PPhy

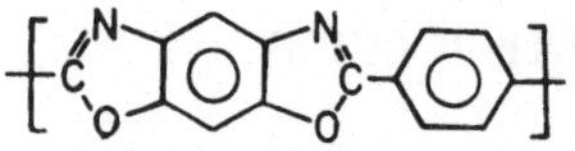

PBO

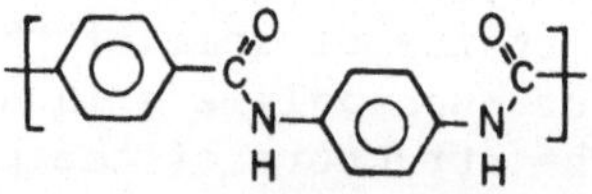

PPTA

The reader is referred to references 1-12 for details on sample preparation and experimental procedures.

Studies on dilute solutions can provide information on intramolecular conformation and intermolecular interactions. The parameters to be considered here include the second virial coefficient A_2, the mean-square radius of gyration $<s^2>$, the chain anisotropy δ, the intrinsic viscosity $[\eta]$, the retention volume V_e in exclusion chromatography (GPC), and the fluoresence emission anisotropy r -- these parameters are studied as functions of molecular weight M, and, where possible, solvent. The first three parameters have been determined by light scattering. All of the polymers cited except PPSQ were studied in solutions in strong acids (methane sulfonic acid, chlorosulfonic acid, etc.) in which they are soluble in the protonated form;[9] PPSQ is soluble in a variety of organic reagents.

Rheological studies on concentrated solutions provide information on intermolecular interactions in addition to data that can help to design processing conditions for fabrication techniques involving concentrated solutions, e.g., fiber spinning, film casting, film extrusion, etc. In this report we will be concerned with the steady-state viscosity η_κ as a function of shear rate κ and the recovery function R_κ following steady-state flow at κ: $R_\kappa = \gamma_\kappa^R/\kappa\eta_\kappa$. The limiting value R_0 at small κ is often denoted by the symbol J_e^0. Some data will also be given on the recovery $R_0(\theta)$ as a function of the time of recovery θ after steady state flow at low κ in the limit where linear viscoelastic behavior obtains.

PROPERTIES OF DILUTE SOLUTIONS

1. Light Scattering Parameters

Although a variety of models can be used to interpret the dependence of the parameters A_2, $<s^2>$, δ, and $[\eta]$ on M, it will suffice for our purpose to consider the relations among these parameters according to the wormlike chain model.[8,13a,14,15] The persistence length ρ and the mass per unit length M_L are important parameters in this model. For example,

$$\frac{<s^2>}{M} = \frac{\rho}{3M_L} \, S(Z) \tag{1a}$$

$$S(Z) = 1 - 3Z^{-1} + 6Z^{-2} - 6Z^{-3}[1 - \exp(-Z)] \tag{1b}$$

$$\delta^2 = \delta_0^2 (2/3 \ Z) \ \{1 - (3Z)^{-1}[1 - \exp(-3Z)]\} \tag{2}$$

where $Z = L/\rho = M/M_L\rho$, with L the chain contour length,
and δ_0 is the intrinsic anisotropy $(-1/2 < \delta_0 < 1)$. In
favorable circumstances, both $<s^2>$ and δ can be determined
from light scattering data. For example, the vertical and
horizontal components of light scattered from vertically
polarized incident radiation is[8] (in the limit of infinite
dilution, denoted by the superscript 0):

$$[R_{V_v}(u,\delta)/KMc]^0 = (1 + \frac{4}{5} \delta^2) - \frac{1}{3}[1 - \frac{4}{5} f_1\delta$$

$$+ \frac{4}{7} f_2^2\delta^2]u + \ldots \tag{3}$$

$$[R_{H_v}(u,\delta)/KMc]^0 = \frac{3}{5} \delta^2 - \frac{9}{35} f_3^2\delta^2 u + \ldots \tag{4}$$

Here, $u = <s^2>h^2$, with $h = (4\pi n/\lambda)\sin(\theta/2)$, and f_1, f_2,
f_3 are functions of L/ρ. Calculations with the worm-
like chain model show that f_1, f_2, and f_3 may be approximated by the limiting value unity for rodlike chains for
practical purposes in the determination of M, $<s^2>$, and δ,
even in the limit of large L/ρ (a coil-like chain) -- in
the latter case, for example, δ is so small that the affect
of putting the f_i equal to unity is negligible.[8]

Data for solutions of BBL in methane sulfonic acid[8]
are shown in Figs. 1 and 2 -- here $<s^2>_{LS}$ denotes the
light scattering averaged $<s^2>$ and N_w is the weight
averaged chain length equal to M_w/m_0, where m_0 is the
repeat unit molar weight ($m_0 = 167$ for BBL). Comparison

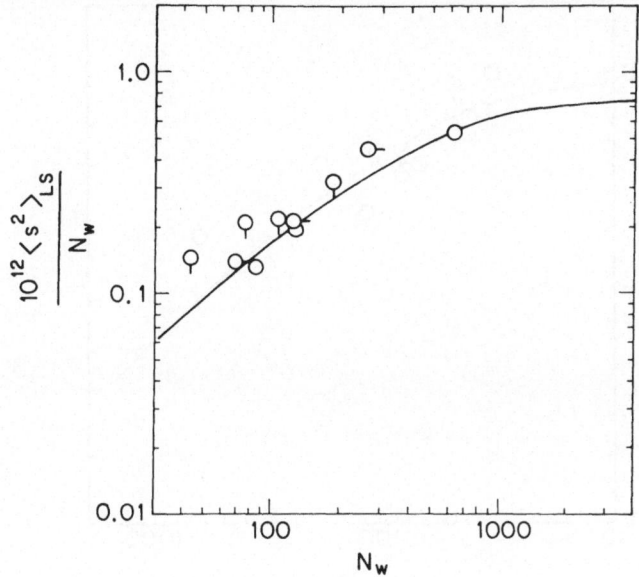

<u>Figure 1.</u> The mean square radius of gyration as a function
of the chain length for fractions of BBL.[8] The curve
represents Eq. (1) with ρ and M_L given in the text.

of the experimental data with Eqs. (1) and (2) yields the
results ρ = 1500 Å and δ_0 = 1 for BBL. By contrast, the
chain coiling that occurs with BBB, PPhy and PPSQ reduces
ρ to 170, 100, and 74 Å for these chains, respectively.[5,6,12]
A model with free rotation about bonds with fixed valence
angles provides a more detailed model for PPhy, leading to
the conclusion that <u>para</u>-catenation predominates for the
bonding of the central phenylene residue in the PPhy repeat unit, giving rise to the large persistence length of
the polymer.[5]

Even though one might expect δ_0 to be large for chains
such as BBB and PPhy, the measured overall chain anisotropy
δ is small (< 0.01), in contrast with δ for BBL, PBO and
PPTA. For the latter two chains, δ in chlorosulfonic acid

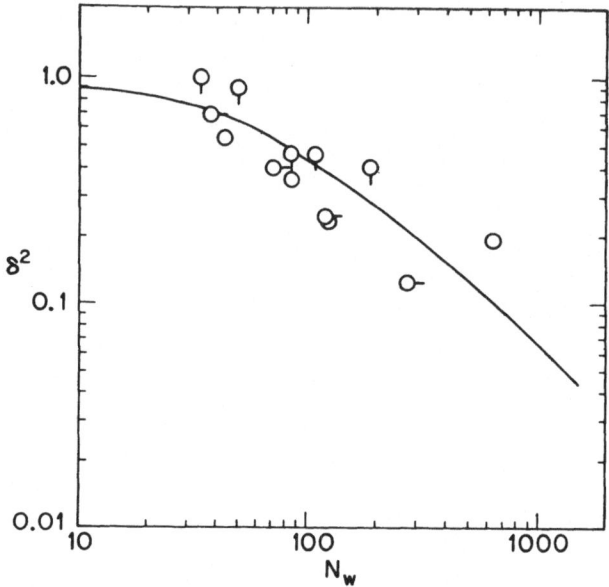

Figure 2. The chain anisotropy as a function of chain length for fractions of BBL.[8] The curve represents Eq. (2) with ρ, M_L and δ_0 given in the text.

is in about unity.[7] With BBB and PPhy, the small value of δ can be attributed to the modest persistence length of the chains and consequent departure from a rodlike conformation, in full accord with the behavior predicted with Eq. (2).

Although meaningful intramolecular properties such as ρ and δ can be determined from data on dilute solutions, caution must be exercised in such interpretations owing to the tendency for intermolecular association in these polymers. For example, metastable aggregates can be formed with fractions of BBL by certain treatments. These aggregates will not dissociate completely in any of the solvents for BBL, and may have a molecular weight many-fold larger than the true value. Some data illustrating this effect are shown in Fig. 3. In these experiments, fractions with a given V_e, prepared by exclusion chromatography, were treated in different ways, and then M_w was determined

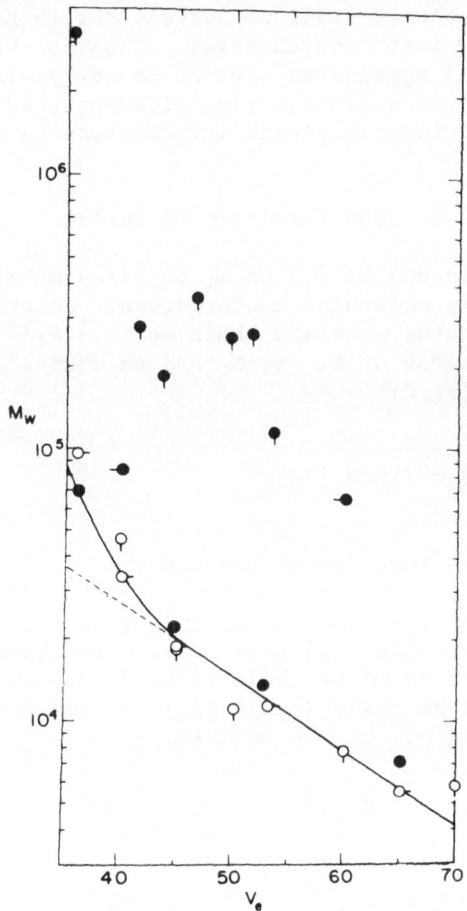

Figure 3. The molecular weight (from light scattering) versus elution volume for fractions of BBL, showing severe effects of association for some of the data which deviate markedly from the solid curve.[8]

with the resultant material. Without discussing the details (which are given elsewhere[8]) it can be seen that some treatments do indeed result in aggregates for which M_w is far larger than the true molecular value. In addition, for these materials δ is very small, usually less than 0.1. This behavior is attributed to aggregation in which the rodlike BBL chains are not packed in a parallel array

(since then δ should still be large), but rather, are associated in some less ordered array. Thus, although measurements with such aggregated systems do not yield values of
δ and ρ with intramolecular significance, they can provide
information on intermolecular interactions in some cases.

2. The Intrinsic Viscosity

The dependence of $[\eta]$ on M_w provides an indirect
method to study molecular conformation. According to calculations with the wormlike chain model,[13a,15] $[\eta]$ can be
expressed in terms of L, ρ, M_L and an equivalent hydrodynamic diameter d:

$$[\eta] = [\pi N_A d^2 / 100 M_L] [\eta]_R \qquad (5)$$

where $[\eta]_R$ is a function of L/d and ρ/d.

For example, estimates for $[\eta]_R$ given by Yamakawa and
Fujii with a wormlike cylinder model[15] and Elzner and
Ptitsyn[16] for a wormlike chain of beads are shown in Fig.
4. For a rodlike chain (small d/ρ) the relation for $[\eta]$
can be approximated by the equation

$$[\eta] \simeq 4.86 \times 10^{20} \frac{d^{0.2}}{M_L} \left(\frac{M}{M_L} \right)^{1.8} \qquad (6)$$

with L/d = $M/M_L d$ in the range 30 to 10^3. It can be seen
that in this limit, $[\eta]$ is only weakly dependent on d, a
result that will be important in some following remarks.

Data on $[\eta]$ as a function of molecular weight for
fractions of PPSQ in two different solvents,[6] are shown in
Fig. 5. For one of these, ethylene dichloride at 50°C,
$A_2 = 0$. With the other solvent, benzene, A_2 is positive,
and of the magnitude commonly observed with polymer solutions in "good solvents". It can be seen that for PPSQ,
$[\eta]$ does not depend very much on solvent, that is on A_2.
Graphical comparison of the data in Fig. 5 with the theoretical curves in Fig. 4 gives ρ = 74 Å in both solvents,
with d equal to 8.9 and 4.4 Å in benzene and ethylene
dichloride, respectively. In this treatment M_L was taken

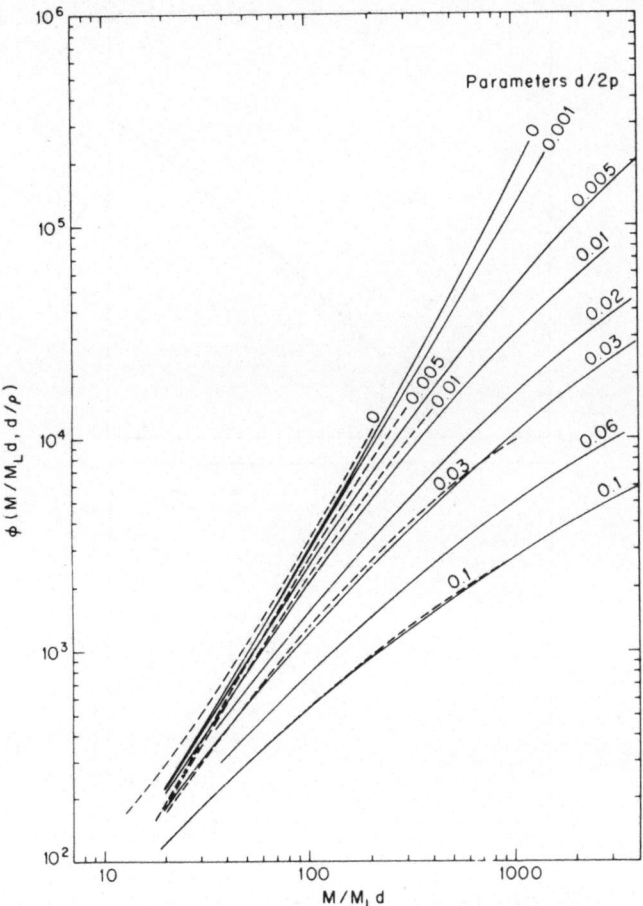

Figure 4. The function ϕ versus M/M_Ld, where $[\eta]_R = K\phi$.
The dashed curves give ϕ according to Eizner and Ptitsyn,[16]
with K = 0.0473, and the solid curves give ϕ according to
Yamakawa and Gujii,[15] with K = 0.0417.

to be 103.2 dalton/Å, in accord with x-ray diffraction
data on PPSQ.[17] The graphical analysis is readily accomplished by comparison of log–log plots of $M_L[\eta]$ versus
M/M_L with the theoretical plots of $[\eta]_R$ versus M/M_Ld.[8]
The value of ρ determined from the viscometric data on
PPSQ is in good accord with the estimate of ρ from light
scattering data.[6]

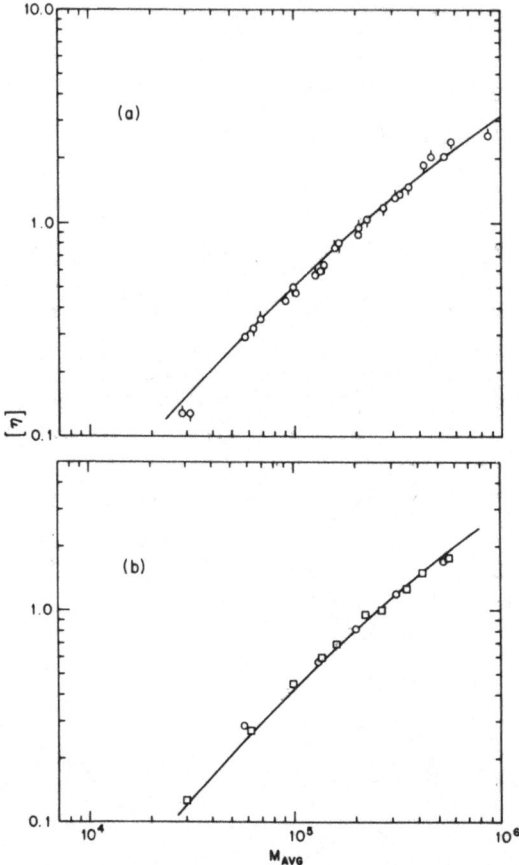

Figure 5. The intrinsic viscosity versus molecular weight for fractions of PPSQ in a) benzene at 25°C, and b) ethylene dichloride at 50.5°C.[6] The curves represent Eq. (5) calculated with values of M_L, ρ and d given in the text.

Similar treatments with data on BBL,[8] PPhy[5] and BBB[12] show that BBL can be treated as a rodlike polymer with d = 10 Å, whereas both PPhy and BBB are coil-like, owing to their small persistence lengths, and in accord with the light scattering estimates of ρ.

It should be noted that nothing in the analysis given above provides information on the intramolecular rigidity of the chains studied. For example, PPSQ has a long persistence length of 75 Å in both a theta solvent (for which $A_2 = 0$) and a solvent with a positive A_2. That is, $\langle s^2 \rangle$ is nearly independent of A_2 for PPSQ. This does not necessarily mean, however, that chain expansion has been prevented by intramolecular rigidity. Thus, the expansion factor $\alpha^2 = \langle s^2 \rangle / \langle s^2 \rangle_0$ can be approximated by the relation[13b,18]

$$\alpha^2 = 1 + a_1 A_2 M^2 / 4\pi^{3/2} N_a \langle s^2 \rangle_0^{3/2} \tag{7}$$

where $\langle s^2 \rangle_0$ is the value of $\langle s^2 \rangle$ when $A_2 = 0$. Here a_1 is a coefficient that is 134/105 for a flexible chain polymer, but is expected to decrease to zero for an intramolecularly rigid chain.[19] Even with the maximum value of 134/105 for a_1, the chain dimension of PPSQ is so large that intramolecular excluded volume effects would be negligible even if the chain is completely flexible. Similar considerations apply with BBB and PPhy.

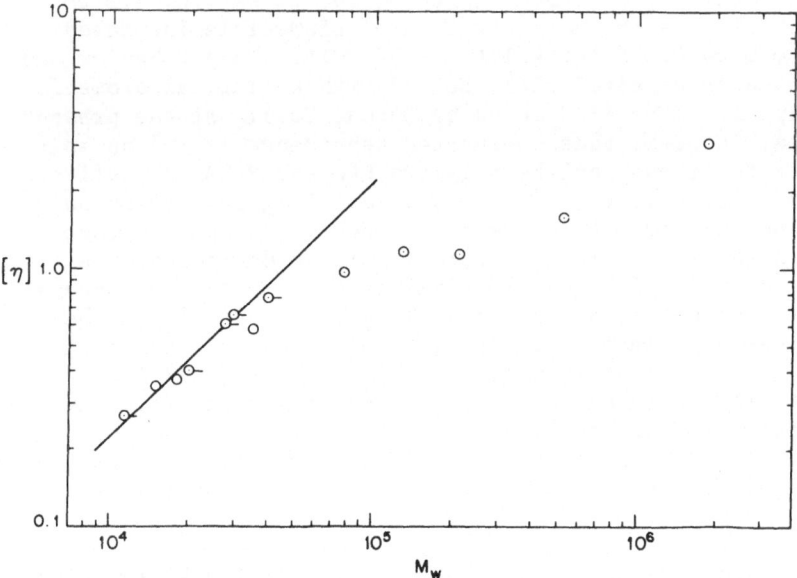

Figure 6. The intrinsic viscosity versus molecular weight for fractions of PPhy, showing deviations due to intermolecular association at large M.[5]

The intermolecular association mentioned above can
also manifest itself in the $[\eta]$-M_w relation. For example,
it has so far not been possible to prevent association in
fractions of PPhy with a molecular weight greater than
about 4×10^4. The effect of this association on the
graph of $[\eta]$ versus M_w for this polymer is shown in Fig. 6.
The line, which corresponds to Eq. (3) with d = 7 Å and
ρ = 100 Å, deviates markedly from the data at higher M,
which are appreciably affected by interchain aggregation.
Similar affects have been observed with all of the polymers
discussed here, except for PPSQ.

A special form of interchain aggregation appears to
occur with rodlike chains under some circumstances. For
example, one would not expect $[\eta]$ to depend on solvent for
PBO or PPTA (to the extent that the latter adopts a rodlike conformation). It is known, however that these
chains exhibit a wide variation of $[\eta]$ in solvents such as
chlorosulfonic acid, methane sulfonic acid, and sulfuric
acid, and that the presence of lithium chlorosulfonate
can affect $[\eta]$ measured in chlorosulfonic acid.[7] Thus,
the data in Fig. 7 show that $[\eta]$ decreases from 4.6 to
1.69 dl/g as the concentration of $LiSO_3Cl$ is increased
from 0 to 0.1 N for solutions of PPTA. This behavior has
also been observed with PBO and with Kevlar, an aromatic
polyamide (supplied to us by DuPont Co.). At the present
time, it seems that unexpected dependence of $[\eta]$ on solvent for these rodlike polymers PBO and PPTA may reflect
association of the chains in parallel aggregates in some
circumstances. If there are ν such chains per aggregate,
then the intrinsic viscosity should be decreased by a
factor ν (neglecting the small effect due to the increased
chain diameter, see Eq. 6). This interpretation is based
on the postulate that aggregation of the rodlike chains
in a parallel array will not much alter the rod length L,
but will decrease the number of effective particles by the
factor ν. Consequently, if ν varies with the polymer
concentration c, or the solvent, so will $[\eta]$. An estimate
of $[\eta]$ at a given concentration c is given by the relation

$$\{2(\eta_{sp} - \ln \eta_{rel})/c^2\}^{1/2} = [\eta] - (1/3 - k')[\eta]^2 c$$

$$+ \ldots \qquad (7)$$

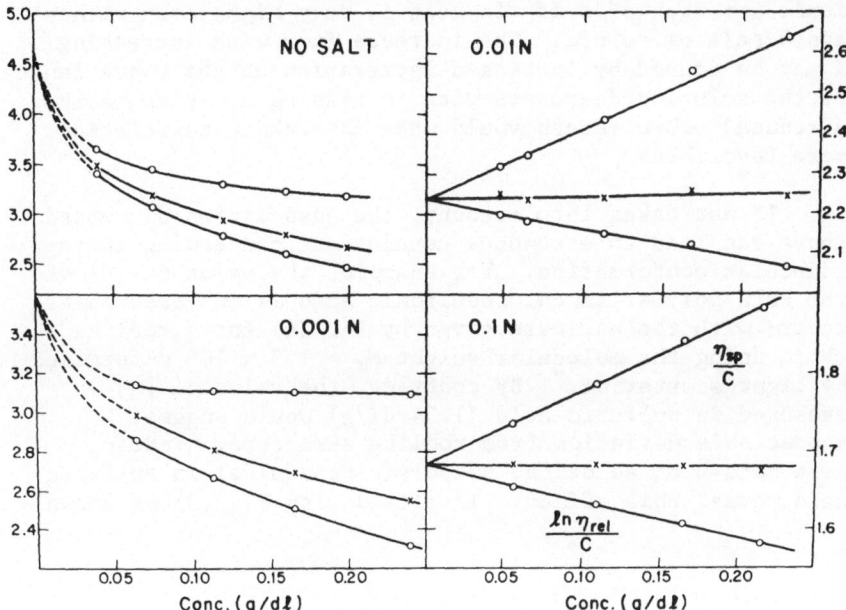

Figure 7. Viscometric data for solutions of PPTA in chlorosulfonic acid containing the indicates concentrations of lithium chlorosulfonate.[7] The crosses are calculated with Eq. (7). The circles are η_{sp}/c or ln η_{rel}/c, as indicated.

Since k' is about 1/3, the second term on the right hand wide of Eq. (7) is usually small enough to neglect. The crosses in Fig. 7, calculated with Eq. (7) with the neglect of this term, show how [η] varies with c for the polymer studied. It can be seen that [η] increases as the ionic strength I of the solution decreases. Contributions to I come from ions produced in the protonation of the polymer by the acid,[9] the added salt, and ions present in the pure acid (from autoprotonation, etc., etc.). Thus, I varies appreciably with c if the salt concentration and the ionic strength of the solvent are low. Conversely, there will be little change in I if the ionic strength from these sources is high. For example, the

ionic strength of sulfuric acid is very high, even without
added salt or solute. The increase in ν with increasing
I may be caused by increased aggregation as the Debye length
of the solution decreases with increasing I. Presumably,
a reduced Debye length would make interchain association
more favorable.

If not taken into account, the association discussed
above can lead to erroneous conclusions concerning intramolecular conformation. For example, the value of [η] of
the PPTA polymer in chlorosulfonic acid is in reasonable
accord with the estimate given by Eq. (4) for a rodlike
chain using the molecular weight $M_w = 1.3 \times 10^4$ determined
by light scattering.[7] By contrast, the value of [η]
measured in sulfuric acid (1.34 dl/g) would suggest
appreciable deviation from rodlike structure. Indeed,
some data[20] on solutions of poly(p-benzamide) in sulfuric
acid reveal this effect. If fitted with Eq. (3) as shown

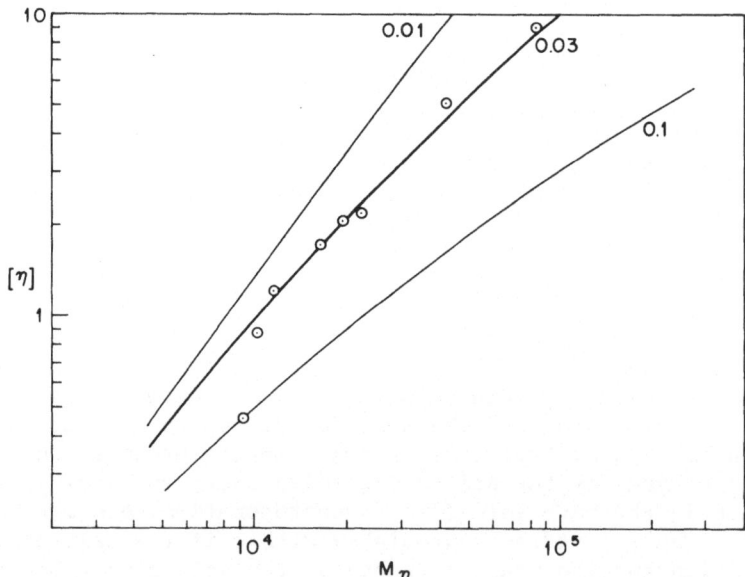

Figure 8. The intrinsic viscosity for PBA in sulfuric
acid.[20] The curves represent Eq. (5) calculated with M_L
and d as given in the text for the indicated values of $\rho/2d$.

in Fig. 8, the data give d = 4 $\overset{\circ}{A}$ and ρ = 70 $\overset{\circ}{A}$, indicating
departure from a rodlike chain conformation. This conclu-
sion is probably fallacious, however, reflecting the
tendency of the polyamide to aggregate in sulfuric acid
rather than a marked deviation from a rodlike conformation.

3. Fluoresence Emission Anisotropy

Several of the polymers studied exhibit fluoresence
emission, and the anisotropy r given by

$$r = \frac{I_{||} - I_{+}}{I_{||} + 2I_{+}} \tag{8}$$

is easily measured. Here $I_{||}$ and I_{+} are the emissions
intensities between parallel and crossed polars, respectively. For rigid particles r is related to the rotatory
diffusion constant D_R:[21]

$$r = r_0(1 + 6D_R\tau_E)^{-1} \tag{9}$$

where τ_E is the lifetime of the excited state (usually
1-10 ns) and r_0 is the intrinsic anisotropy, equal to
2/5 if the absorption and emission vectors are parallel,
as would be expected for an electronic transition along
the axis of a rodlike molecule. In general,[13c]

$$D_R = KRT/100\eta_0M[\eta] , \tag{10}$$

with $[\eta]$ in dl/g, where K is a numerical constant and η_0
is the solvent viscosity. For example, K is 2/15 and 1/4
for rigid rod and coil, respectively. Combination of
Eqs. (9) and (10) suggests that a plot of r^{-1} versus
$\tau_E RT/\eta_0[\eta]M$ should be linear, with intercept r_0^{-1} and
slope $6K/100r_0$. The data on BBL shown in Fig. 9 exhibit
the anticipated linearity, but the slope is 100 times
longer than expected for a rodlike chain. The probable
source of this discrepancy is the effect of molecular
weight polydispersity on the measurement.[8] This is quite

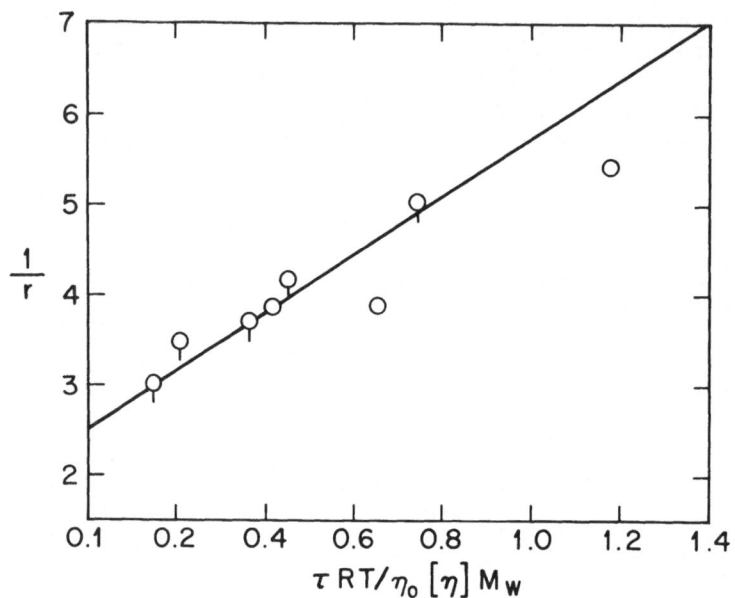

Figure 9. The reciprocal fluoresence emission anisotropy versus $\tau_E RT/\eta_0[\eta]M_w$ for fractions of BBL.[8]

severe, since for a rodlike chain D_R is nearly proportional to M^{-3}. Consequently in a plot of $\bar{r}^{-1}$ versus $\tau_E RT/\eta_0[\eta]M_w$ for a polydisperse polymer the slope should be $6K(M_w/M_R)/100r_0$, where (M_w/M_R) is given by

$$M_w/M_R = M_z M_w^{2} M_n \int_0^\infty M^{-4} f(M) dM \tag{11}$$

with $f(M)$ the weight density of M. Thus, the measurement is too sensitive to polydispersity to be of much value in determining intramolecular structure.

With PPhy, the situation is a little different. The internal modes of motion cannot be neglected and if a relation of the form of Eq. (7) is used, D_R must be

interpreted as a properly conformation averaged quantity.
Similar problems arise in the interpretation of certain
spin-lattice relaxation times determined in nuclear magnetic relaxation experiments.[22] An adequate theory for
this analysis is not now available. The correlation of
$\bar{r}^{-1}$ with $\tau_E RT/\eta_0 [\eta] M_w$ for fractions of different M_w shown
in Fig. 10 does suggest, however, that long-range motions
of the chain are important. It may suffice, for example,
to consider only the first few longest relaxation times in
a normal mode analysis of the problem.

RHEOLOGICAL PROPERTIES OF CONCENTRATED SOLUTIONS

Calculations[23-25] of the steady state viscosity η_K
as a function of the shear rate κ for (isolated) rodlike
chains can be put in the form

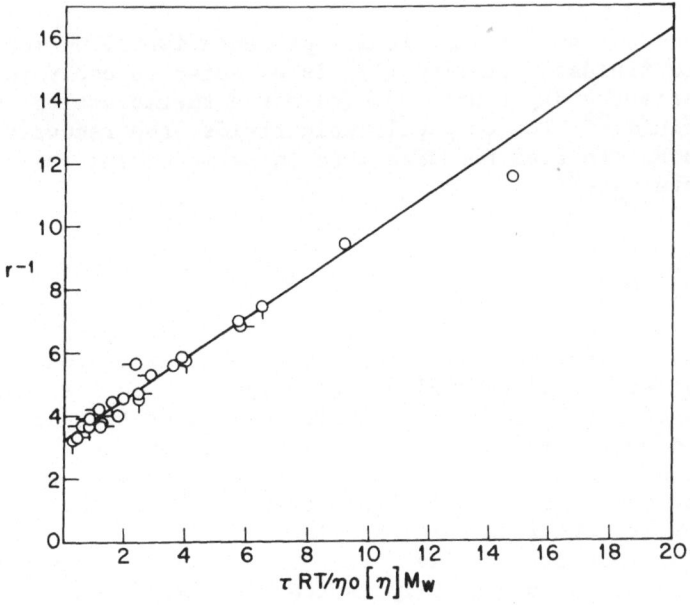

Figure 10. The reciprocal emission anisotropy versus
$\tau_E RT/\eta_0 M_w$ for fractions of PPhy.[5]

$$\eta_\kappa = \eta_0 \{1 - C_1(\tau_c\kappa)^2 + C_2(\tau_c\kappa)^4 - \dots\} \qquad (12)$$

where η_0 is the limiting value of η_κ at small κ, and the characteristic time τ_c is defined as

$$\tau_c = \eta_0 R_0$$

Here, R_0 is the value of the steady state recovery function $R_\kappa = \gamma_\kappa^R/\kappa\eta_\kappa$ in the limit of small κ, with γ_κ^R the recovery after cessation of steady state creep at shear rate κ. The constants C_1 and C_2 are equal to 0.7286 and 0.1996, respectively, for rodlike chains. Equation (12) is a special form of the relation[26,27]

$$\eta_\kappa = \eta_0 Q(\tau_c\kappa) \qquad (13)$$

that has been used in empirical representation of η_κ for polymeric fluids. Equation (13) is expected to apply for the temperature dependence of $\eta(\kappa)$ for a thermorheologically simple fluid.[28] For many polymeric fluids, the recovery function R_κ can also be correlated in terms of the reduced shear rate $\tau_c\kappa$[27]:

$$R_\kappa = R_0 P(\tau_c\kappa) \qquad (14)$$

In the following, we will discuss the behavior of concentrated solutions of several of the polymers described above in terms of these relations. It is well known that concentrated solutions of rodlike chains will undergo a phase transition from a disordered to an ordered state at some molecular weight dependence concentration.[29] In the ordered state the rods form parallel arrays, usually leading to optical anisotropy. The first part of our discussion will be confined to optically isotropic, disordered solutions. In the final parts of this section we will turn to some properties of ordered optically anisotropic solutions and optically isotropic, aggregated solutions.

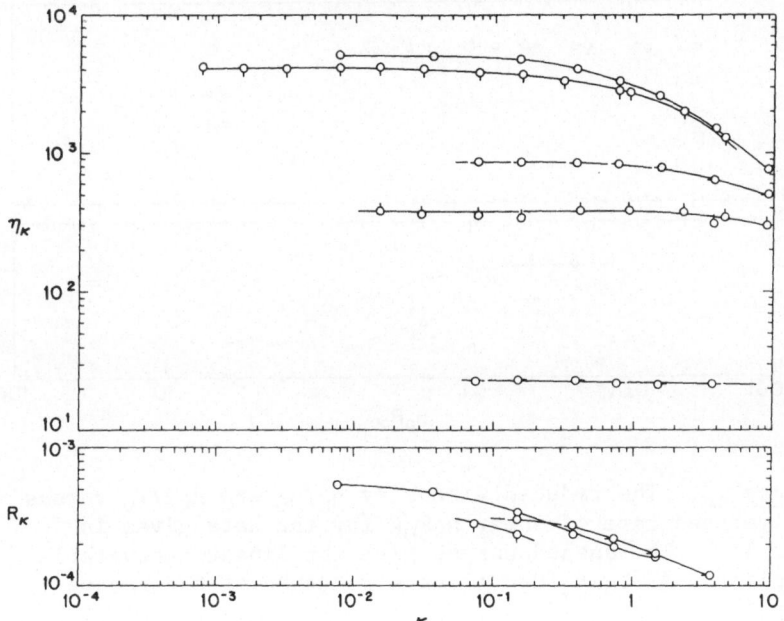

$\underline{\text{Figure 11}}$. The viscosity and the recovery function versus
shear rate for solutions of three fractions of PPSQ in
cyclohexanone (0.2 g/ml) at several temperatures.[6]

1. Isotropic Solutions

Experimental details of the determination of η_K, R_K,
and other rheological properties discussed below, may
be found elsewhere.[6,7,10,30] Suffice it to say that
measurements were made with a specially constructed cone
and plate rheometer. The platens are fabricated from
gold to permit the use of solutions in strong acids.

Data on optically isotropic solutions of each of the
polymers studied could be correlated with Eqs. (13)-(14),
provided extensive intermolecular aggregation was not a
problem (see below). For example, data on concentrated
solutions of PPSQ,[6] BBB[10] and Kevlar[7] are shown in Figs.
11-14. It can be seen that for each polymer η_K begins to

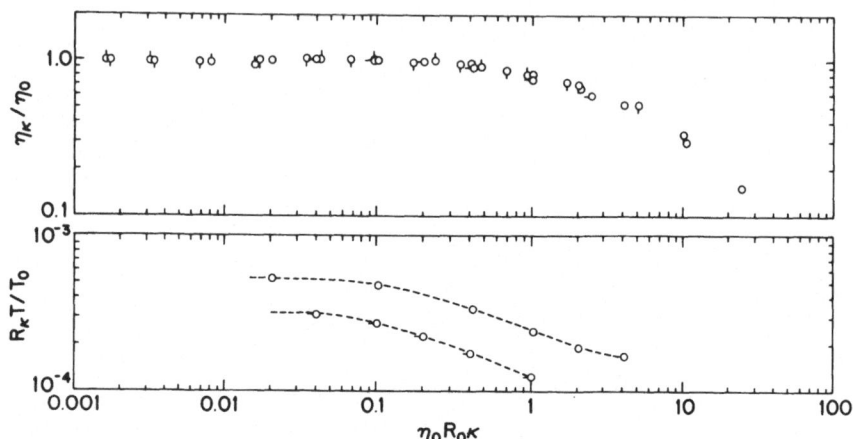

<u>Figure 12.</u> The reduced viscosity η_κ/η_0 and $R_\kappa T/T_0$ versus
the reduced rate of shear $\eta_0 R_0 \kappa$ for the data given in
Fig. 11.[6] The dashed curves give the linear recoverable
compliance $(T/T_0)R_0(\theta = \kappa^{-1})$ as a function of the recovery
time θ.

deviate markedly from η_0 when $\tau_c \kappa$ is about 0.7. This does
not imply that the polymers have similar conformations
in concentrated solutions. Indeed, parallel behavior has
been observed with a whole range of flexible chain polymers.[27] What can be concluded is that the reciprocal time
constant τ_c^{-1} computed from the <u>linear viscoelastic</u>
constants η_0 and R_0 provides a good measure of the deformation rate for which nonlinear phenomena occur with
polymeric fluids independently of the specifics of the
chain conformation. For example, with isolated rodlike
chains, τ_c in Eq. (12) is $(3/5)\tau_R$, where τ_R is the time
constant $(2D_R)^{-1}$ for rotatory diffusion.[23] For a flexible-
coil chain polymer with a spectrum $H(\tau)$ of relaxation
times, $\tau_c = \tau_0 R_0$ can be identified as the average

$$\tau_c = \int_\infty^\infty \tau^2 H(\tau) d\ln\tau \bigg/ \int_\infty^\infty \tau H(\tau) d\ln\tau \tag{15}$$

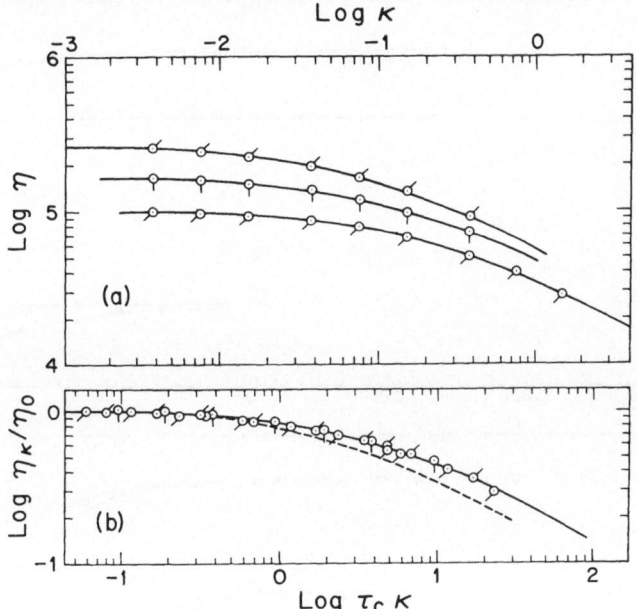

Figure 13. (a) The viscosity versus the shear rate for solutions of BBB in methane sulfonic acid at three temperatures.[7] (b) The reduced viscosity versus the reduced shear rate for the data in part (a). The dashed curve shows similar data for polyisobutylene solution for comparison.

(Of course, Eq. (15) can also be applied to rodlike polymers). For a rigid coil, one might again expect proportionality between τ_C and τ_R. Whatever the molecular interpretation in a particular case, it appears that τ_C is a useful phenomenological parameter for predicting the onset of nonlinear viscous behavior.

Similar remarks apply to the correlation of R_κ/R_0 with $\tau_C\kappa$. The dependence of R_κ/R_0 on temperature is accounted for satisfactorily by the reduced shear rate, just as with flexible chain polymers. It should also be noted that deviation from linear behavior (e.g., from $R_\kappa \sim R_0$) is

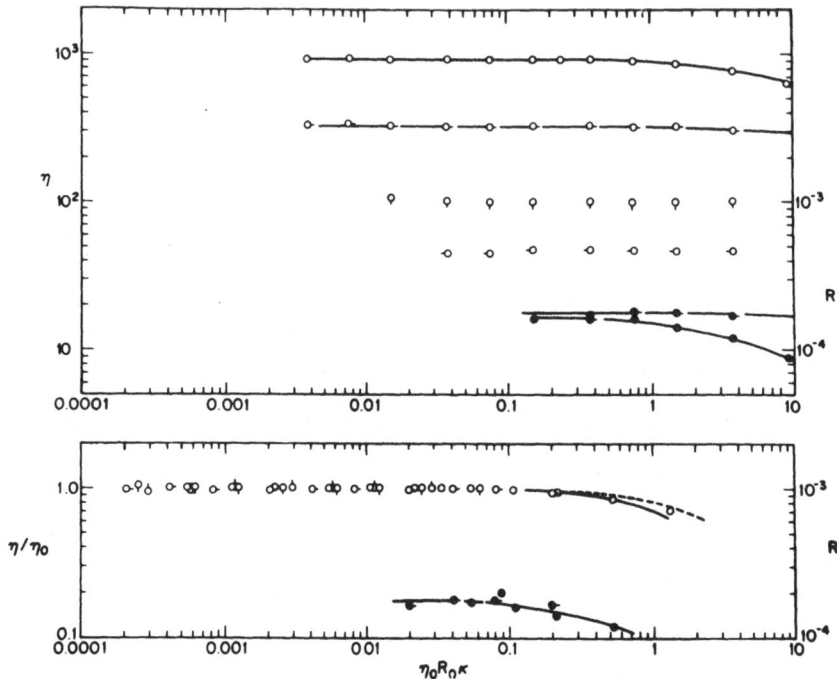

<u>Figure 14</u>. <u>Upper</u>: The viscosity and the recovery function
versus shear rate for an optically isotropic solution of
Kevlar at several temperatures from 8 to 72°C.[7] <u>Lower</u>:
The reduced viscosity and the recovery function versus
the reduced rate of shear for the data in the upper figure.

apparent in a range of $\tau_c\kappa$ for which η_κ is still equal to
η_0 within experimental error. In studies on concentrated
solutions of poly(α-methyl styrene) it has been found that
the recovery $R_0(\theta)$ as a function of the time of recovery
θ following cessation of steady-state flow in the linear
viscoelastic range is similar to R_κ measured in the nonlinear range.[27] That is, R_κ and $R_0(\theta = \kappa^{-1})$ will superpose
over a limited range of κ, usually for $\tau_c\kappa$ less than about
10. Similar behavior is observed with solutions of PPSQ,[6]
as may be seen in Fig. 12 where the dashed lines give R_0
as a function of $\theta/\tau_c = (\tau_c\kappa)^{-1}$. Since $R_0(\theta)$ can be represented in terms of a spectrum $L(\tau)$ of retardation times
τ in the form

$$R_0(\theta) = J_g + \int_\infty^\infty L(\tau)[1 - \exp(-\theta/\tau)]d\ln\tau \tag{16}$$

with J_g the 'instantaneous compliance', the observed similarity of R_κ and $R_0(\theta = \kappa^{-1})$ suggests that for small enough $\tau_c\kappa$ (e.g., $\tau_c\kappa$ less than about 10) the nonlinearity in the steady-state recovery function with increasing κ may result from a reduced emphasis on molecular motions with long τ. Thus, for $\tau_c\kappa$ less than about 10, R_κ can be represented with the approximate relation

$$R_\kappa \approx J_g + \int_\infty^\infty L(\tau)[1 - \exp(-1/\kappa\tau)]d\ln\tau \tag{17}$$

in which $L(\tau)$ is weighted by the function $1-\exp(-1/\kappa\tau)$. In any case, it does not appear that the dependence of R_κ on $\tau_c\kappa$ for the polymers studied here differs markedly from the properties of flexible chain polymers reported elsewhere.

The dependence of η_0 and R_0 on polymer concentration c and molecular weight M provides another means to compare the properties of the polymers studied here with the more familiar behavior of flexible chain polymers. For the latter, for example, the viscosity of concentrated solutions is given by the relation[31]

$$\eta_0 = (N_A/6) \times (cM/\rho M_c)^a\zeta \tag{18}$$

where

$$a = \begin{cases} 2.4 & \text{if } X > X_c \\ 0 & \text{if } X \le X_c \end{cases}$$

$$X = c<s^2>_0/m_a = cM(<s^2>_0/Mm_a)$$

Here, m_a and ζ are the molar weight and friction factor per hydrodynamic unit, respectively, ρ is the polymer density, M_c is a polymer specific constant, and $X_c = \rho M_c (< s^2 >_0 / M m_a)$. For many flexible chain polymers X_c is about 400×10^{-17} if m_a is defined as the molar weight per chain atom, and the friction factor is given by

$$\zeta = \zeta_0 \exp \frac{C}{T - T_g + \Delta} \tag{19}$$

where neither C or Δ depend much on c. Thus, if $T_g - \Delta$ is much greater than T_g, ζ will not depend appreciably on c, and at constant T one should have $\eta_0 \propto (cM)^{3.4}$ if $cM \geq \rho M_c$, and $\eta_0 \propto cM$ if $cM \leq \rho M_c$.

The data in Fig. 15 for concentrated solutions of PPSQ in two solvents, benzene and cyclohexanone, show that similar behavior is observed with this polymer,[6] despite the presumed intramolecular rigidity of the chain. Similar results have been reported for BBB.[3,12] The measured values of $< s^2 >_0 / M$, and ρM_c lead to the result $m_a X_c = 2.30 \times 10^{-13}$ for PPSQ. If the repeat unit is assumed to have four hydrodynamic units, one for each chain atom along the "rails" of the chain skeleton, then $X_c = 360 \times 10^{-17}$ for PPSQ, an estimate very close to the 'universal' value of 400×10^{-17} observed with a variety of flexible chain polymers. Consequently, for PPSQ, the dependence of η_0 on c and M does not reflect the presumed rigidity of the molecule in any marked way in the range of these parameters so far reported.

Finally, we may inquire about the dependence of R_0, or τ_c, on c and M. For many (monodisperse) flexible chain polymers R_0 can be expressed in the form[26,27,32]

$$c^2 (R_0)_{flex} = \frac{2}{5} \frac{E^* \rho M_e}{RT} \tag{20}$$

if the entanglement density $E = c\rho / \rho M_e$ exceeds a critical value E^* of about 10. Here M_e is the molecular weight between entanglement loci (M_e is often about $M_c/2$). According to Eq. (20), R_0 should be independent of M (in the range $E > 10 E^*$). The data in Fig. 12 shows that this is not the case for fractions of PPSQ. In addition, if it is assumed

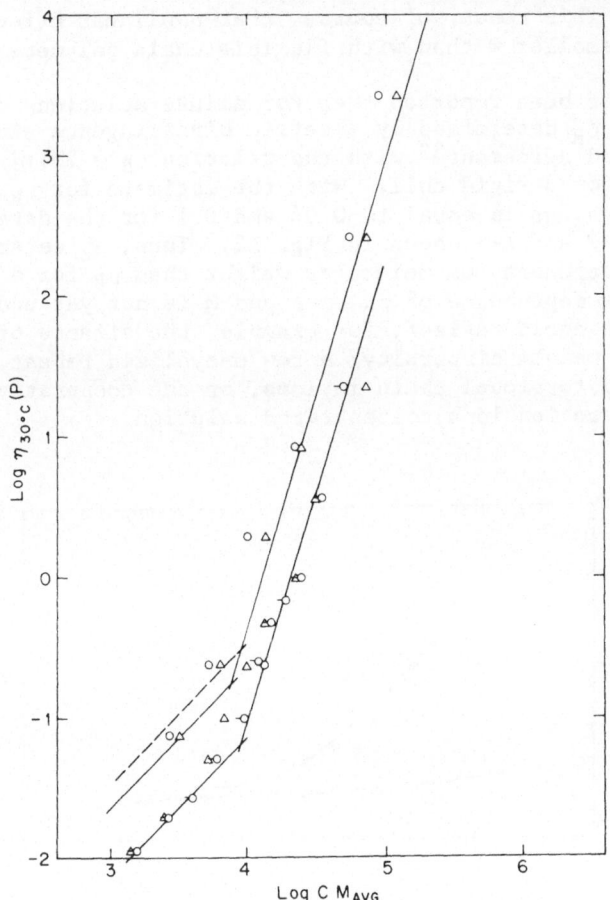

<u>Figure 15</u>. The viscosity versus the product cM_{AVG} for
solutions of PPSQ in benzene, lower curve, or cyclohexanone,
upper curve.[6] M_{AVG} is either M_w or $1.5M_n \sim M_w$.

that $E^*M_e = 5M_c$, as would usually be true with monodisperse
flexible chains,[26] then according to Eq. (20) one should
have $(R_0)_{flex} = 0.2 \times 10^{-4}$ cm^2/dyne for the 0.2 g/ml PPSQ
s-lutions, compared with the experimental values which lie
in the range 3–5 $\times 10^{-4}$ cm^2/dyne. Thus, for PPSQ the
characteristic time $\tau_c = \eta_0 R_0$ is both longer, and more
dependent on M than would be expected for a flexible chain

polymer. This means, of course, that nonlinear effects
occur at smaller κ than with flexible chain polymers.

It has been reported that for dilute solutions of PPSQ
values of τ_R determined by electric birefringence studies
are in good agreement[33] with the relation $\tau_R = 2M[\eta]\eta/RT$
expected for a rigid coil. With the estimate for τ_R, we
find that τ_C/τ_R is equal to 0.04 and 0.1 for the data on
samples A-2 and A-2 shown in Fig. 12. Thus, τ_C is smaller,
and less dependent on molecular weight than τ_R for a rigid
coil. The dependence of τ_C on c and M is not yet understood. It could reflect, for example, the effects of
molecular weight dispersity, a few uncyclized repeat units,
restricted torsional chain motions, or the cooperative
nature of motion in a concentrated solution.

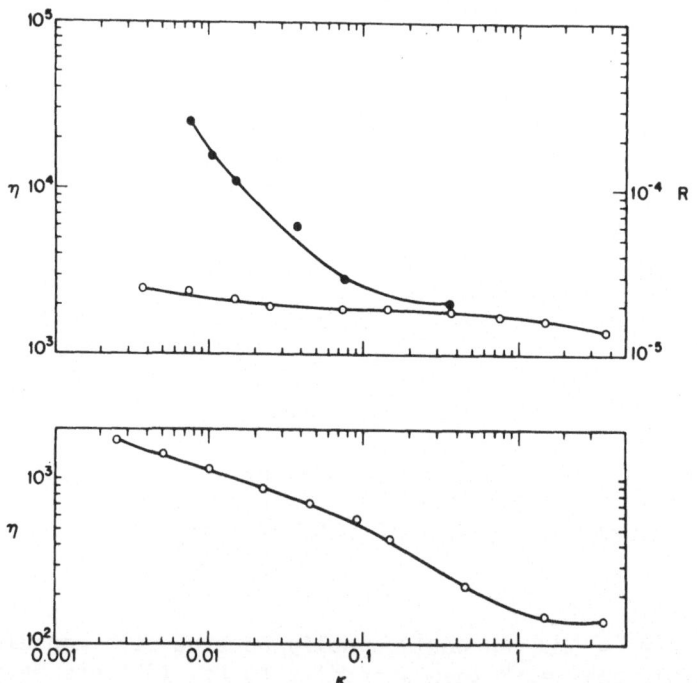

Figure 16. Upper: Viscosity (○) and recovery function
(●) versus shear rate for an optically isotropic PBO
solution. Lower: Viscosity versus shear rate for a slightly
aggregated optically isotropic PBO solution.[7]

2. Anisotropic Solutions

Flow curves for solutions of Kevlar and PBO just concentrated enough to exhibit optical anisotropy are shown
in Figs. 16 and 17.[7] Two effects completely absent in the
data for isotropic solutions may be noted: the viscosity
η_K (calculated as the shear stress σ divided by the steady
state rate of strain κ) and the recovery function R_K both
increase with decreasing κ at low κ. These effects can be
understood more clearly by reference to the reduced curve
for the data on Kevlar solutions shown in Fig. 18.   In
Fig. 18, η_{REF} is equal to η_0 for the isotropic solution, but
for the anisotropic solution η_{REF} is put equal to η_K in
the range of κ for which η_K levels off after decreasing
(e.g., in the range with $\tau_c\kappa$ about 0.5). Similarly, τ_c is
computed from η_0 and the measured R_0 for the isotropic
solutions, but for the anisotropic solution τ_c is taken as
the empirical constant to force superposition of the data
on isotropic and anisotropic solutions in the range
$\tau_c\kappa \approx 1$. The τ_c so deduced is similar to the value one
would obtain by extrapolation of τ_c versus c from data on
isotropic solutions. The behavior of η_K for the anisotropic
solution in the range $\tau_c\kappa < 0.1$ is attributed to the nematic
liquid crystalline structure in the quiescent fluid. For
shear rates less than about $0.1/\tau_c$, this structure is maintained, and during deformation, the shear stress is nearly
independent of the rate of strain. Apparently, the structure is able to rupture and reform rapidly in comparison
with the applied rate of deformation if $\kappa < 0.1/\tau_c$, so that
σ is closely related to the stress required to induce
slippage in the ordered nematic structure.

The effect on R_K shown in Fig. 16 is typical of our
observations with optically anisotropic solutions of rodlike polymers -- R_K appears to be both smaller and more
dependent on κ than would usually be expected for an isotropic solution. Indeed, if c is much greater than the
critical concentration required for optical anisotropy,
$\gamma_K{}^R$ is so small as to be immeasurable.

It is evident from the behavior displayed in Figs. 16-
18 that a plot of "the viscosity" versus concentration can
take on various forms, depending on the range of $\tau_c\kappa$ for
which η_K is measured. For example, in a plot of η_0 versus
c, the viscosity would diverge to infinity for c in the
range for optical anisotropy.

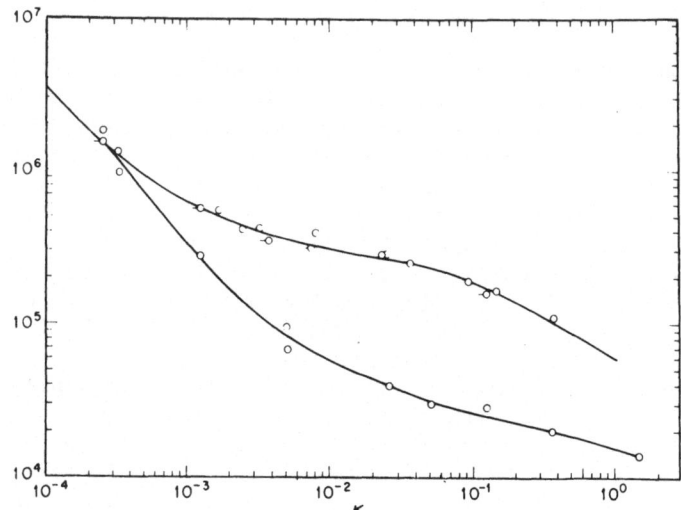

<u>Figure 17.</u> The viscosity versus shear rate for an optically
anisotropic solution of Kevlar at 52.2°C, lower, and 8.8°C,
upper curve.[7]

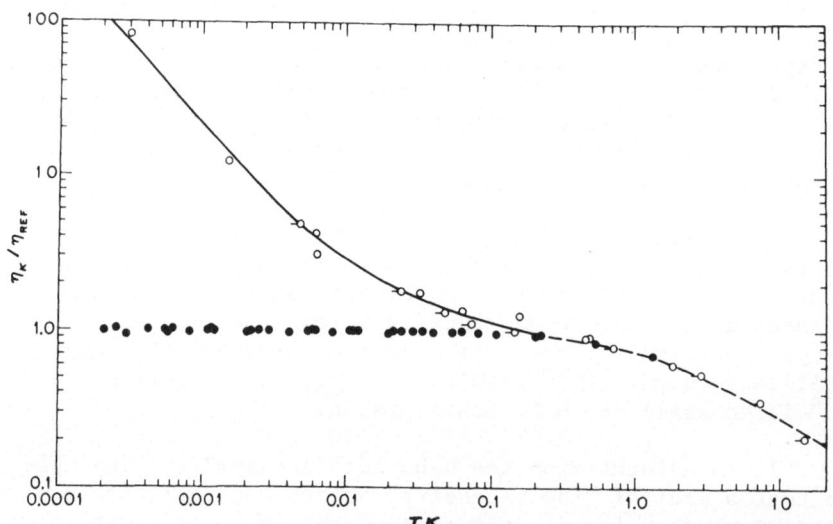

<u>Figure 18.</u> The reduced viscosity versus the reduced shear
rate for the data in Figs. 14 (●) and 17 (O).

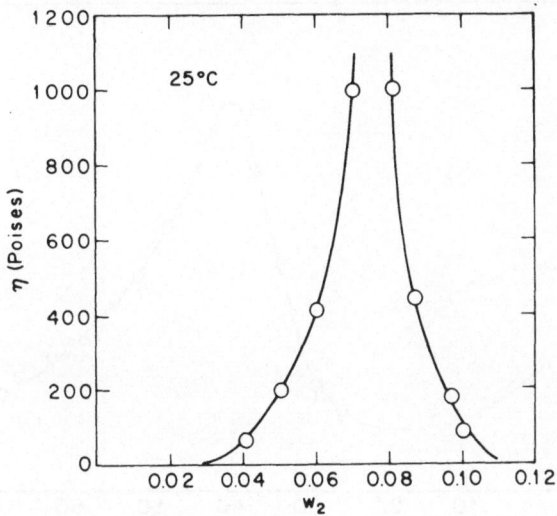

<u>Figure 19.</u> The viscosity versus concentration for solutions
of PBO.7 See text for a discussion of the conditions for
determination of η.

A more instructive plot can be obtained by examining
η_{REF} (e.g. η_K) for $\tau_c K$ in the range 0.1 to 1 versus c. A
plot of such data for solutions of PBO is shown in Fig. 19.
The extremum in the resultant η_K versus c curve is similar
to that reported with solutions of PPTA,[34] poly(p-benzamide)[35,36] and poly(γ-benzyl-L-glutamate).[37] Indeed,
the extremum has been used to determine the concentration
c* for the onset of the phase transition from the disordered
to the ordered state.

In addition to the apparently anomalous dependence of
η_{REF} on c displayed in Fig. 19, one also finds unusual
behavior for the dependence of η_{REF} on temperature. Data
on a PBO solution exhibited in Fig. 20 show that η_{REF}
first increases with decreasing T, as is usual, but below
a temperature T*, η_{REF} decreases with decreasing T until a
minimum is reached at T**, with η_{REF} again increasing with
decreasing T at still lower T. This complex temperature
dependence reflects the onset of phase separation at T* and

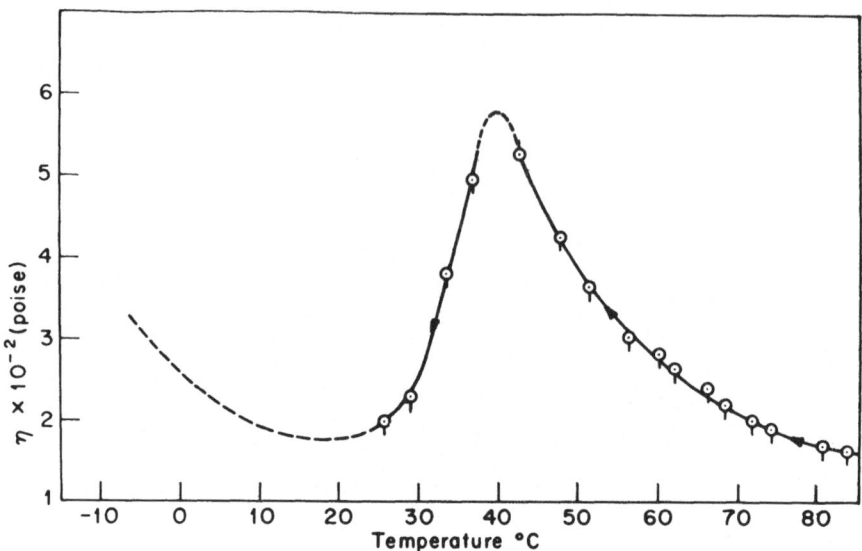

<u>Figure 20.</u> The viscosity versus temperature for a solution
of PBO showing the temperature (ca. 40°) of transformation
to an optically anisotropic solution.

the disappearance of the final isotropic regions at T^{**}.
The wide temperature interval of the phase transition
probably reflects the broad molecular weight distribution
of the PBO polymer used, although it may also reflect the
practical difficulty of achieving equilibrium conditions
in the experiment. For example, the data for η_{REF} versus
T in Fig. 20 for T increasing and decreasing do not superpose exactly, as might be expected if equilibrium conditions
prevailed in the transition from the ordered to the disordered state.

The data in Figs. 19 and 20 on PBO solutions are similar
to data reported by Papkov and coworkers on PPTA and
poly(p-benzamide) solutions.[34,36] A schematic representation summarizing the behavior observed in our laboratory
and by Papkov and coworkers for η_{REF} as a function of T and
c for rodlike polymers that can undergo a order-disorder

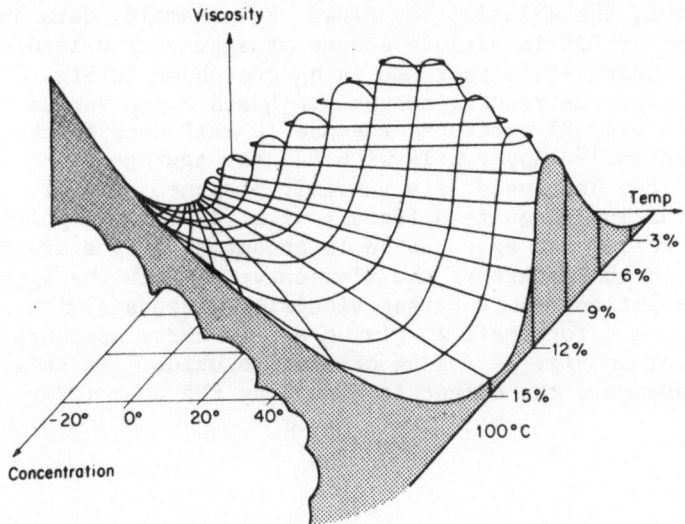

Figure 21. Schematic diagram of the dependence of the viscosity on temperature and concentration for polymers such as PBO, PBA and PPTA. Based on data in references 34 and 36.

phase transition is illustrated in Fig. 21. The general characteristics of this dependence will probably apply to many rodlike polymers.

3. Aggregated Solutions

In the work discussed above, the polymers PPTA, BBL, BBB and PBO were in solution in some strong acid (e.g., methane sulfonic acid, chlorosulfonic acid, etc.). Solubility of these polymers in a strong acid involves protonation of the polymer, and solvation of the consequent polyelectrolyte. The introduction of a small (e.g., 1%) amount of water can cause deprotonation and aggregation of the polymer. This aggregation can, for example, prevent the order-disorder phase transition that might be expected for an anhydrous solution in the case of rodlike polymers. If the aggregation is not too

extensive, the solution may flow. For example, data on a
solution of BBB in various stages of aggregation leading
to one hundred-fold increase in η_0 are shown in Fig. 22.
Remarkably, the reduced curves η_κ/η_0 and R_κ/R_0 versus $\tau_c\kappa$
shown in Fig. 23 superpose reasonably well despite the
aggregation.[10] Apparently with BBB the aggregated
species can be viewed as a randomly branched polymer.
The situation is quite different with the rodlike polymer
BBL in which case aggregation is accompanied by a drastic
change in the nature of the flow curve.[7] With the aggregated solutions the apparent viscosity increases with
decreasing κ for small κ, giving a flow curve somewhat
like that in Fig. 17 for an ordered solution. In this
case, however, the effect is caused by the weak inter-

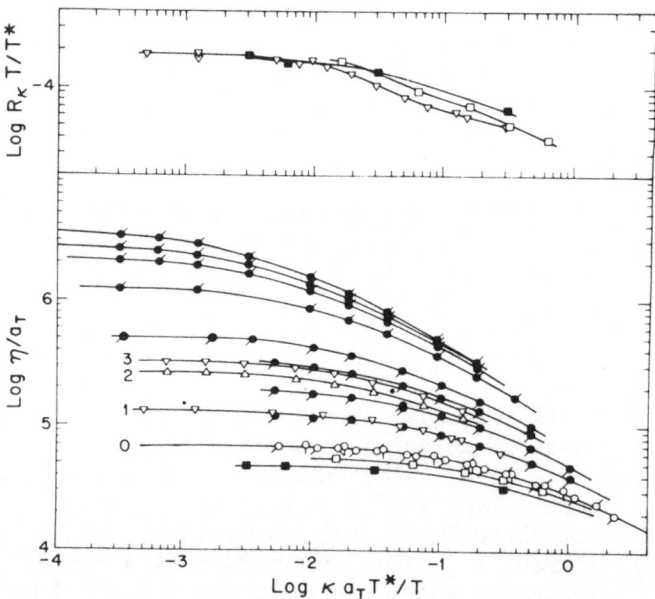

Figure 22. The viscosity and the recovery function versus
shear rate for solutions of BBB undergoing aggregation.[7]
The factors a_T correct the data to a constant temperature.

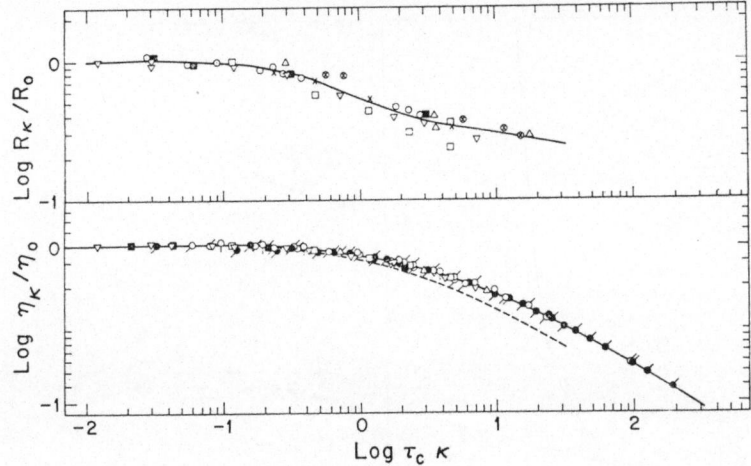

<u>Figure 23</u>. The reduced viscosity and recovery function versus the reduced rate of shear for the data in Fig. 22.

molecular interactions causing association of the rodlike molecules into aggregates in which the molecules do not appear to be parallel.

If the association is extensive enough, a weak solid is formed that will exhibit a yield point. The data in Fig. 24 illustrate the properties observed with such an aggregated solution. For strains less than a critical value γ_y of about 5% the creep was fully recoverable and fitted by the cube-root, or Andrade, creep relation for the compliance $\gamma(t)/\sigma$:[7]

$$J(t) = J_A[1 + (t/\tau_A)^{1/3}] \tag{21}$$

In this case, the recovery function $R(\theta)$ for recovery as a function of the time θ following creep of duration S is given by

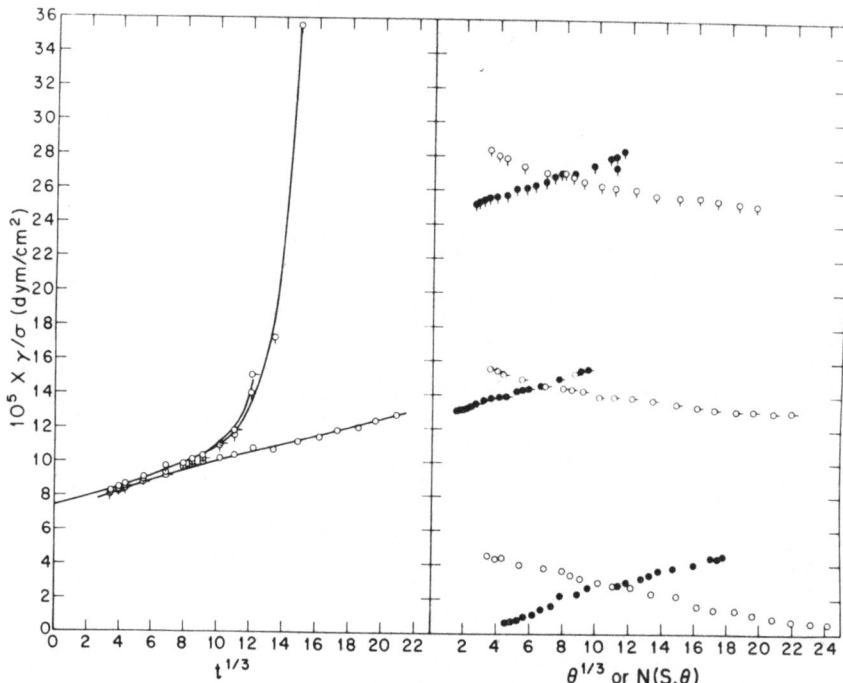

Figure 24. Left: Creep as a function of the cube-root of time for an aggregated solution of BBL at three different stress levels: ○ , 331.7; ○- , 456.3; and ○ 563.4 dyn/cm². Right: The recovery versus the cube-root of the time of recovery θ, ○ , or the function N(S, θ), ● , given by Eq. (22c).

$$R(\theta) = J(\theta + S) - J(\theta) \qquad\qquad (22a)$$

$$R(\theta) = J_A \tau_A^{-1/3} N(\theta, S) \qquad\qquad (22b)$$

$$N(\theta, S) = (\theta + S)^{1/3} - \theta^{1/3} \qquad\qquad (22c)$$

For the data in Fig. 24, τ_A and J_A are about 2×10^4 sec and 7.4×10^{-5} cm²/dyne, respectively. Thus, even though the aggregated solution has a rather low initial modulus

$J_A^{-1} = 1.4 \times 10^4$, recoverable creep takes place on a very slow time scale. This creep may involve deformation within the supramolecular aggregates of the rodlike BBL molecules, perhaps with some reorientation of the molecular stacking in the aggregate. Similar cube-root creep has been observed with BBB and a variety of other polymers.[1]

In experiments for which the strain $\gamma(t)$ exceeds γ_y, a term μt must be added to Eq. (21) to account for the nonrecoverable deformation occuring for $\gamma(t) > \gamma_y$. The effect of the term is seen in Fig. 24 as a rapid increase in γ/σ at about 1000 sec. Remarkably, the recovery is still given by Eq. (21a) (or by Eq. (22b) if a term about equal to μS is added to account for the nonrecoverable deformation). Apparently the strain within the aggregates is maintained during the period of nonrecoverable strain, and is recovered once the stress is removed. The non-recoverable deformation may involve relative motion of neighboring supramolecular aggregates. This model for the deformation of the aggregated BBL solution is reinforced by the observation that whereas J_A is very dependent on temperature γ_y is only weakly so. The properties of weakly aggregated solutions of PBO and PPTA are qualitatively similar to those just described for BBL -- it requires different concentrations of moisture to reach an equivalent state of aggregation with each polymer.

The overall behavior of rodlike polymers discussed above can have important consequences in processing techniques which utilize concentrated solutions of rodlike polymers.

For example, in fiber spinning of a concentrated solution of a rodlike polymer it will be necessary to have extensional flow at rates in excess of about $10 \ \tau_c^{-1}$ in order to achieve the desired molecular orientation in the final fiber. Moreover, the orientation so induced in the concentrated solution must be quenched in a time short in comparison with τ_c by the removal of the solvent. This process is aided considerably by the use of solutions sufficiently concentrated to be in the ordered state. On the other hand, intermolecular interactions which act to trap the system in the disordered state by aggregations may make it impossible to achieve the desired orientation in the fiber.

ACKNOWLEDGEMENT

It is a pleasure to acknowledge support from the National Science Foundation, Division of Solid State Chemistry and Polymer Science, and Wright-Patterson Air Force Base, Nonmetallic Materials Division, for parts of the work described above.

REFERENCES

1. G. C. Berry, J. Polym. Sci., Polym. Phys. Ed., 14, 451 (1976).

2. G. C. Berry and M. Murakami, J. Polym. Sci., Polym. Phys. Ed., 14, 1721 (1976).

3. G. C. Berry and S. M. Liwak, J. Polym. Sci., Polym. Phys. Ed., 14, 1717 (1976).

4. C.-P. Wong and G. C. Berry, ACS Polym. Prepr., 17, (2), 413 (1976).

5. J. L. Work and E. F. Casassa, ACS Polym. Prepr., 18 (1), 000 (1977).

6. T. E. Helminiak and G. C. Berry, ACS Polym. Prepr., 18(1), 000 (1977).

7. C.-P. Wong, H. Ohnuma and G. C. Berry, ACS Polym. Prepr., 18(1), 000 (1977).

8. G. C. Berry, ACS Polym. Prepr., 18 (1), 000 (1977).

9. G. C. Berry and P. R. Eisaman, J. Polym. Sci., Polym. Phys. Ed., 12, 2253 (1974).

10. C.-P. Wong and G. C. Berry, ACS Org. Coatings Plas. Chem. Prepr., 33(1), 215 (1973).

11. V. G. Ammons and G. C. Berry, J. Polym. Sci., Pt. A-2, 10, 449 (1972).

12. G. C. Berry, Disc. Faraday Soc., 49, 121 (1970).

13. H. Yamakawa, Theory of Polymer Solutions, Harper and Row, New York (1971), a) p. 56; b) p.

14. K. Nagai, Polym. J., 3, 67 (1972).

15. H. Yamakawa and M. Fujii, Macromolecules, 7, 128 (1974).

16. Y. E. Eizner and O. B. Ptitsyn, Vysokolmol. Soedin, 4, 1725 (1962).

17. J. F. Brown, Jr., J. Polymer Sci., Pt. C, 1, 83 (1963).

18. G. C. Berry and E. F. Casassa, J. Polym. Sci., Pt. D, 4, 1 (1970).

19. H. Yamakawa and W. H. Stockmayer, J. Chem. Phys., 57, 2843 (1972).

20. J. R. Schaefgen, V. S. Foldi, F. M. Logullo, V. H. Good, L. W. Gulrich, and F. L. Killian, ACS Polym. Prepr. 17(1), 69 (1976).

21. D. J. R. Laurence, in Physical Methods in Macromolecular Chemistry, Ed., B. Carroll, Marcel Dekker, New York, 1969, Chap. 5.

22. A. A. Jones and W. H. Stockmayer, J. Polym. Sci., Polym. Phys. Ed., 00, 000 (1977).

23. J. G. Kirkwood and R. J. Plock, J. Chem. Phys., 24, 665 (1956).

24. E. Paul, J. Chem. Phys., 51, 1271 (1965).

25. T. Kotaka, J. Chem. Phys., 30, 1556 (1959).

26. W. W. Graessley, Adv. Polym. Sci., 16, 1 (1974).

27. G. C. Berry, B. L. Hager and C.-P. Wong, Macromolecules, 10(1), 000 (1977).

28. H. Markovitz, J. Polym. Sci., Polym. Symp. 50, 431 (1975).

29. P. J. Flory, Proc. Royal Soc. (London), A234,
    73 (1956).

30. G. C. Berry and C.-P. Wong, J. Polym. Sci., Polym.
    Phys. Ed., 13, 1761 (1975).

31. G. C. Berry and T. G Fox, Adv. Polym. Sci., 5, 231
    (1968).

32. E. Riande, H. Markovitz, D. J. Plazek and N.
    Raghupathi, J. Polym. Sci., Polym. Symp. 50, 405
    (1975).

33. V. N. Tsvetkov, K. A. Andrianov, Ye. N. Ryumtsev,
    I. N. Shtennikova, M. G. Vitovskaya, N. N.
    Makarova and N. A. Kurasheva, Polym. J. (USSR),
    15, 455 (1973).

34. T. S. Sokolova, S. G. Yefimova, A. V. Volokhina,
    G. I. Kudryavtsev and S. P. Papkov, Polym. Sci.
    (USSR) 15, 2832 (1973).

35. S. L. Kwolek, P. W. Morgan, J. R. Schaefgen and
    L. W. Gulrich, ACS Polym. Prepr., 17(1), 53 (1976).

36. S. P. Papkov, V. G. Kulichikhin, V. C. Kalmykova,
    and A. Ya. Malkin, J. Polym. Sci., Polym. Phys. Ed.,
    12, 1753 (1974).

37. J. Hermans, Jr., J. Colloid Sci., 17, 638 (1962).

DISCUSSION

P. FLORY - STANFORD U. Just a small comment on your invoking the entropy of solution. It should be the entropy of disorientation. We agree that it is the stiffness of the chain that really matters. But once the molecules are disoriented, entropy contributions associated with chain stiffness have been dissipated. G. BERRY - CARNEGIE-MELLON UNIVERSITY. I agree. I think our experience in attempting to redissolve precipitated and dried fractions point to this also. I don't understand why the fractions are more difficult to dissolve than the unfractionated material, but if you take an unfractionated material and a fraction, precipitate them in the same way, and then try to put each into solution, in some cases the fraction will not go into solution at all in the same solvent in which the unfractionated material can be dissolved very easily. I think it is the order that is important.

P. FLORY. I should like to comment further on that very point, a most interesting observation. We started some calculations, i.e., Dr. Abe started some calculations about six months ago (neither he nor I have been able to finish them) on heterodisperse systems and there are marked effects of polydispersity. Although this work is incomplete, it is evident that such effects are substantial. One can see from theory they are very large. Hence, I would like to believe your results. Finally, on the matter of recovery, wouldn't you expect that if you had a macroscopic liquid crystal domain and it was oriented, its reorientation time would be exceedingly long? G. BERRY. I believe that is correct. One of the slides that I showed for the recovery for a solution with a concentration a little less than that where anisotropy develops shows that the time constant is very long. In fact, we were unable to measure the recovery at concentrations well into the anisotropic range. Nonetheless, I think that if you want to make a highly oriented material you have to worry about even small amounts of disorder developing after flow has ceased, and these kinds of considerations must be remembered when one tries to design process conditions to fabricate ordered materials from concentrated solutions.

R. G. SINCLAIR - BATTELLE MEMORIAL INSTITUTE. Isn't
it possible that interchain branching could produce some of
the scatter we saw in some of the early plots of BBL?
G. BERRY. I don't think so, because the polymers are
identical. It's not that we change from one polymer to
another during that series of experiments, but rather, we
look at one material under various conditions of precipitation and dissolution.

R. G. SINCLAIR. I'm referring, e.g., to such things
as the intrinsic viscosity plot vs. molecular weight where
there were some black dots, i.e., what you attribute to
aggregation, out of line. I agree that aggregation may
be a dominant cause of the nonlinearity, but it should be
kept in mind that interchain branching will cause the same
scatter. Ladder polymers such as BBL invariably have such
aberrant structure. G. BERRY. The slide giving $[\eta]$ vs.
M_W for BBL solutions had some data (e.g., the filled circles)
for aggregated BBL sample; they were well off the expected
correlation. In fact, those data were obtained with a
polymer that when treated differently gave data on the
expected line (e.g., for the open circles). It's not that
we're looking at different polymers, we're looking at the
same polymer under different states of manipulation. I
think branching is one way of describing the kind of intermolecular structure that is formed. I think of it as a
kind of branched structure, but the branching does not arise
from covalent bonds. Instead, the branched structure is
formed by secondary valence interactions.

R. BAUGHMAN - ALLIED CHEMICAL. Suppose that you align
the polymer chains in the liquid crystal phase via shear
between two plates, and then shear these plates in an
orthogonal direction. Do you think that reorientation will
occur initially inhomogeneously as, for instance, in the
plastic deformation of a metal? I ask this question for a
particular reason. G. BERRY. I'm pleased to give an
answer immediately. I don't know. I do know that the
orientational birefringence develops very quickly. You
don't have to do much of anything, as I think Dr. Morgan
also emphasized. My guess is that if you did the experiment you suggested, you would see it shifting over to the
other orientation very quickly.

R. BAUGHMAN. The reason I asked this question is that if the reorientation process occurs inhomogeneously (so as to produce intermediate states in which some planar domains have the initial chain orientation and other planar domains have the orthogonal orientation), then it may be possible to quench in a unique type of biaxial polymer morphology.
G. BERRY. I don't think it would happen, but I don't know.

C. OVERBERGER - U. MICHIGAN. My recollection tells me there is some controversy about that phenyl T ladder. After that polymer was discussed by the General Electric Group, an NMR paper appeared (I can't tell you where it appeared) discussing the structure of this material. Has that been resolved? They claimed it was a branched material, an imperfect structure. I don't remember the details. I can find it but I wonder if that has been clarified as to whether that is really a ladder.   G. BERRY. I don't know the answer to that. I remember the same thing you do and about the same date (about 1969 or 1970). What I can tell you is that, based on more dilute dilution work than I was able to talk about here, we feel that the light scattering radii and intrinsic viscosity can be fitted by a model with a persistence length of about 70 A. I can't say any more than that.

M. SHEN - U. OF CALIFORNIA. Do I understand that the system is thermorheologically simple in that you obtain the master curve by time, temperature reduction?   G. BERRY. That is correct.

M. SHEN. And that the implication would be that the structure of the polymer solution (whatever it happens to be) is not affected by the shear rate.   G. BERRY. No, I wouldn't say that. It means that the frictional forces are about the same over that range of temperature that we studied. The data on "Kevlar" aramid fiber polymer appeared to be thermorheologically simple. That data is taken not very far away from the peak point. I think if you go further (and we haven't done that), it might not be thermorheologically simple. Probably, at higher concentrations the structure will be much stronger, and be affected by temperature in a different way than is the viscosity. Where we show the data, the solutions are rheologically simple.

M. SHEN. Did you compare your dilute solution data
with the Kirkwood theory?   G. BERRY. The equations for
the viscosity of rodlike molecules we have been using are
due to Yamakawa and Fujii or Ptitsyn and Eizner. They are
very close to the relation given by Kirkwood and Auer in
the range of interest to us.

PHASE EQUILIBRIA IN RIGID-CHAIN POLYMER SYSTEMS

S. P. Papkov

Artificial Fiber Institute

Moscow, USSR

ABSTRACT

One of the specific features of these polymer systems is a low mobility of the macromolecules and correspondingly slow phase transition rates. This enables one to use, in analyzing such systems, composite phase diagrams showing all the types of phase equilibria inherent in a given system. Extension of this principle to the systems "rigid-chain polymer-solvent" makes it possible to construct a phase diagram which combines (a) equilibrium with the formation of a crystalline phase, (b) equilibrium with the formation of liquid-crystalline phases, and (c) equilibrium with the formation of amorphous (liquid) phases.

The paper considers a general type of such a composite phase diagram for a rigid-chain polymer and gives examples of phase transitions with delayed kinetics in setting up the equilibrium state.

The author emphasizes the importance of the formation of non-equilibrium states and the kinetics of transition to equilibrium states for cases of processing rigid-chain polymers into fibers and films.

One of the specific features of these polymer systems is a low mobility of the macromolecules and corresponding-ly slow rates of the processes associated with phase trans-itions. This phenomenon is well known for the case of polymer crystallization in bulk and in solution. The rate of nucleation of the crystalline phase is often so slow that in practice one has to deal with the amorphous state of the polymer, although thermodynamic equilibrium corres-ponds to its crystallization.

The present author attempts to extend the phenomenon of superposition of different types of phase equilibrium to the case of systems containing rigid-chain polymers, where not only amorphous and crystalline phases, but also a liquid-crystalline phase, can be formed.

In the first place, it is well to recall the general form of the phase diagrams for each of the types of equi-librium. Figure 1 shows schematically the phase equili-

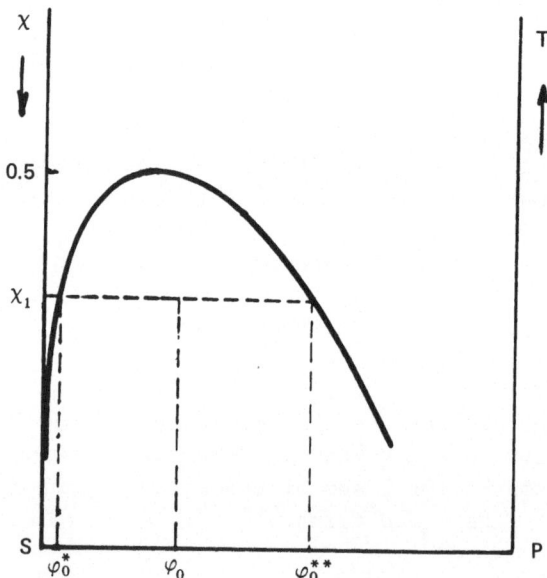

Figure 1.   Phase equilibria in a polymer-solvent system
            with formation of amorphous (liquid) phases.

brium diagram for the case of separation of the system
into two amorphous phases (liquid equilibrium). On attainment of values of Huggins' parameter χ >0.5 (for instance,
χ_1)or, accordingly, the temperature T_1, which lies above
the critical point of compatibility of the polymer and the
solvent, a one-phase solution of a polymer P in a solvent
S separates into equilibrium amorphous phases φ_o* and φ_o**,
where phase φ_o* is practically pure solvent, while phase
φ_o** is a highly concentrated isotropic solution of the
polymer.

Figure 2 demonstrates, in a general form, the phase
equilibrium diagram for a crystallizing polymer. With
changes in parameter χ or in temperature T, the region of
coexistence of the isotropic solution and the crystalline
phase is defined by a curve with an asymptotic approach
to S at low polymer concentrations.

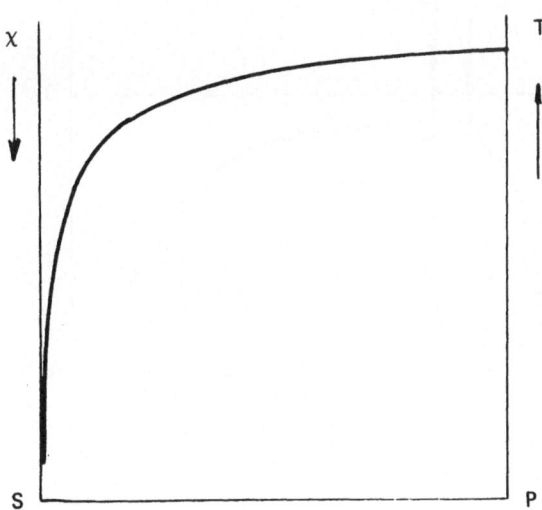

Figure 2.   Phase equilibria in a polymer-solvent system
            with formation of a crystalline·phase.

   The phase equilibrium in systems containing rigid-
chain polymers is characterized by the formation of a
liquid-crystalline state, which fact can be illustrated by
the diagram due to Flory[1] reproduced in Figure 3. At χ
values below 0, the polymer-solvent system forms either
an isotropic (one-phase) solution ($\varphi < \varphi_{\ell c}^{*}$) or a mixture of
an isotropic and anisotropic (liquid-crystalline) phase
($\varphi_{\ell c}^{*} < \varphi < \varphi_{\ell c}^{**}$), or a liquid-crystalline phase of variable
composition ($\varphi > \varphi_{\ell c}^{**}$). At $\chi > 0.07$, the region of coexistence of the isotropic and liquid-crystalline phases
greatly expands. For instance, for χ_1, the composition
of the liquid-crystalline phase is equal to $\varphi_{\ell c_1}^{**}$.

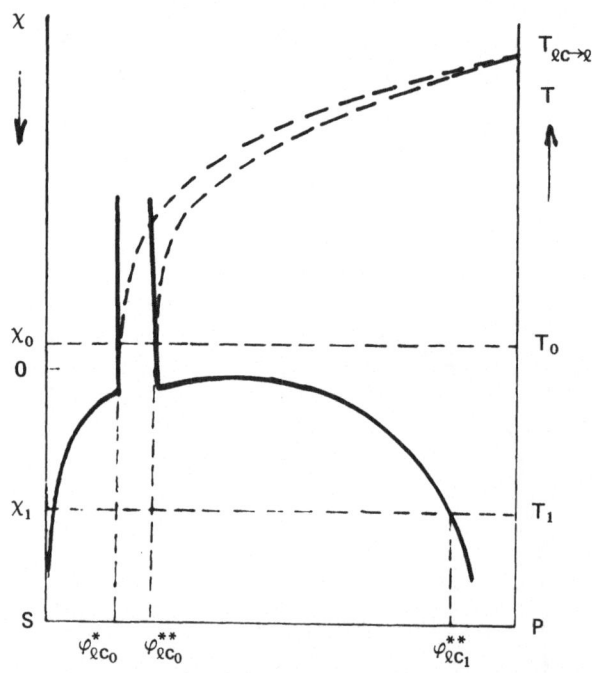

Figure 3. Phase equilibria in a "rigid-chain-polymer-
            solvent" system with formation of a liquid-
            crystalline phase.

At present, it is difficult to draw a conclusion with regard to the change in composition of the coexisting phases in the χ range below 0.07. According to Flory's theoretical calculations the composition of these phases change only slightly with decreasing χ or increasing T. It should be assumed, however, that by analogy with the behavior of low-molecular-weight liquid crystals, there must be a temperature at which, in a pure polymer containing no solvent, transition from the liquid-crystalline to the isotropic state takes place. This transition must occur above the fusion temperature of the cyrstalline polymer (enantiotropic transition). Therefore, curves of compositions $\varphi_{\ell c}^*$ and $\varphi_{\ell c}^{**}$ must, in principle, extrapolate from the position predicted by Flory, as is shown by the dotted lines in Figure 3, where the point $T_{\ell c \to \ell}$ denotes transition of the polymer from the anisotropic to the isotropic state. Experimental data for the system "poly-γ-benzyl-L-glutamate + dimethylformamide"[2] and for the system "poly($\underline{p}$-phenyleneterephthalamide) + sulfuric acid"[3] show a trend toward such a change in the composition of the coexisting phases, as indicated in Fig. 4. Thermal decomposition of most of the rigid-chain polymers begins below their fusion temperature, and this hinders experimental detection of the point $T_{\ell c \to \ell}$.

It is possible to construct, quite independently, a hypothetical phase diagram showing simultaneously all the three types of phase equilibria, i.e., amorphous, crystalline and liquid-crystalline. This superposition of the three types of equilibria, based on the difference in the rates of nucleation of the new phases, is presented schematically in Figure 5.

The main objective pursued in this case is to analyze the phase transitions occurring in real systems "rigid-chain polymer-solvent" with an allowance for the kinetics of these transitions.

Let us choose, as the initial system, a polymer solution with parameters $\chi_o(T_o)$-φ_o. This solution can be

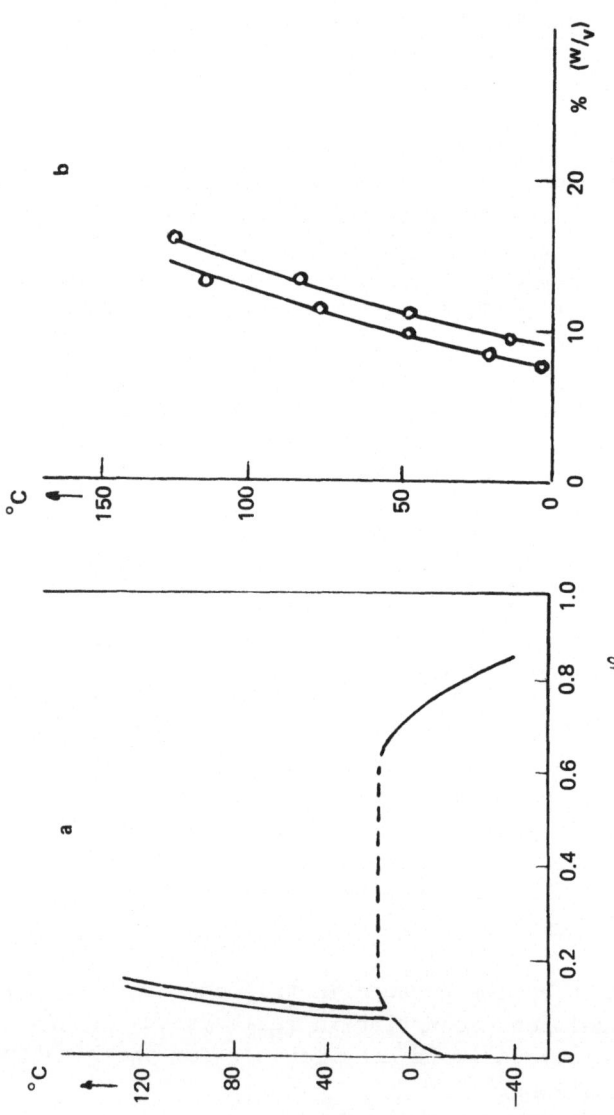

Figure 4.   Phase equilibria with formation of a liquid-crystalline phase for the systems:
a – poly-γ-benzyl-L-glutamate-dimethylformamide; b – poly(p-phenyleneterephthalamide)-sulfuric acid.

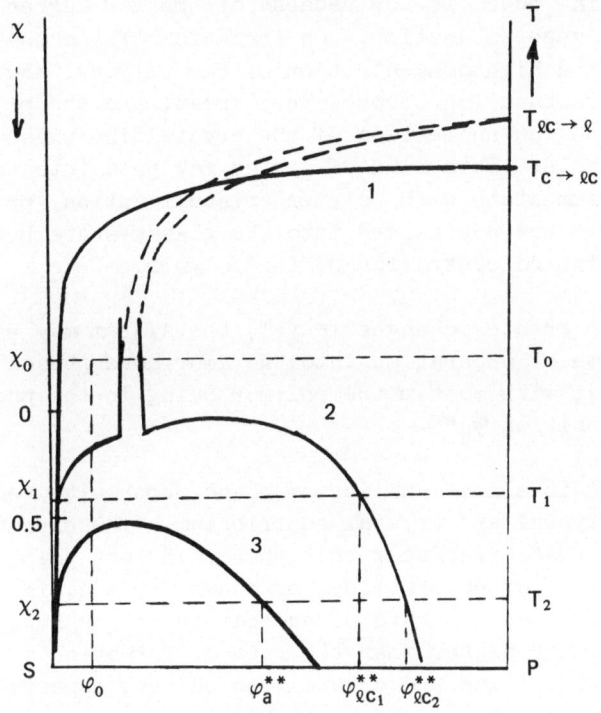

Figure 5.   Generalized phase diagram for a "rigid-chain
            polymer - solvent":1 - crystalline equilibria,
            2 - liquid-crystalline equilibria, 3 - amorphous
            equilibria

obtained by adding a solvent to the polymer, which is in
an amorphous or liquid-crystalline (but not crystalline)
state. With decrease in temperature to T_1 or an increase
in the parameter χ to χ_1 (for instance, as a result of
addition of a non-solvent) the system may pass into the
region where separation into isotropic and liquid crystal-
line ($\varphi_{\ell c_1}^{**}$) phases can occur. In both cases (at χ_0 and χ_1)
the system is located below the crystallization curve.
For $\chi_0 - \varphi_0$, however, the rate of formation of the

crystalline phase is low because of the low degree of
solution supersaturation. On formation of the phase
$\Psi_{\ell c_1}^{**}$ with a high concentration of the polymer, however,
the supersaturation becomes very great, and the probability
of fluctuation nucleation of the crystalline phase becomes
considerable. This kind of system may pass into the
equilibrium state with polymer crystallization, provided
the system has not passed into the glassy state because
of the high concentration of the polymer.

With greater changes in $\chi(T)$ the system may pass into
the region of separation into two amorphous phases (at
χ_2 or T_2), with most of the polymer being in the phase with
the composition Ψ_a^{**}.

This is a non-equilibrium phase, both with respect to
liquid-crystal and crystal equilibrium. The explanation of
the fact that precisely this phase, and not the liquid-
crystalline (or crystalline, of course) phase, is first
formed, is that the rate of nucleation of the amorphous
phase is much higher than the rate of formation of the
liquid-crystalline and crystalline phases, especially in
spinodal separation. The probability of nucleation is
lower, the higher the phase order.

The rate of transition to the equilibrium state, i.e.,
the rate of transition of the amorphous phase to the
liquid-crystalline state, of composition $\Psi_{\ell c_2}^{**}$, or further
into the crystalline state also depends on the viscosity
of the amorphous phase. This transition is accelerated
by increase in temperature due to the resulting higher
mobility of the macromolecules.

The following experiments may serve as examples of the
described type of transition. An isotropic solution of
poly(p-benzamide) in N,N-dimethylacetamide (+LiCl) with a
concentration of one per cent is far below the critical concentration for the formation of the liquid-crystalline
state. Mixing of such a solution with an excess of non-
solvent (acetone) results in the formation of a precipitate

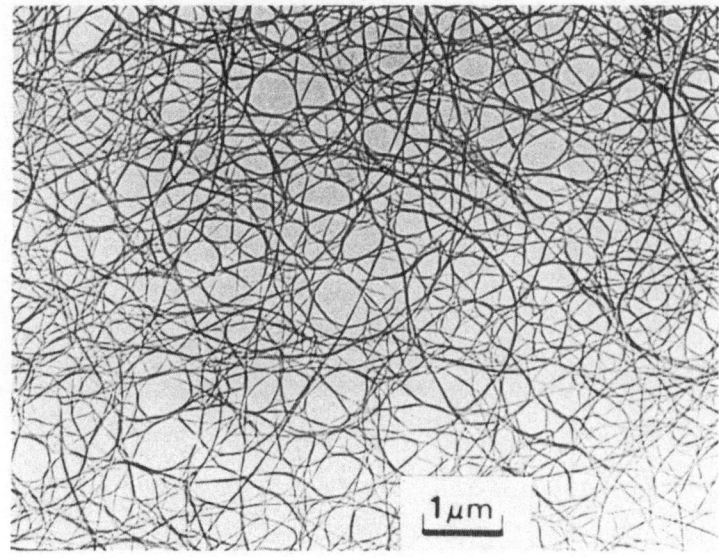

Figure 6. Electron micrograph of a precipitate of poly(p-
benzamide) separated from solution in N,N-di-
methylacetamide by mixing with acetone.

of fibrillar structure, which is characteristic of pre-
cipitates from the liquid-crystalline state. An electron-
micrograph of such a precipitate is presented in Fig. 6.[4]

Another example refers to the transition from the
liquid-crystalline to the crystalline state. While studying the phase equilibrium in the system poly(p-benzamide),
dimethylacetamide (+LiCl), a diagram was obtained which is
reproduced in Fig. 7. The compositions of the coexisting
phases (isotropic and anisotropic) were determined experimentally. While conducting measurements in the region
above 100°C, it was found that on holding the system for
one or two days at this temperature both phases pass
irreversibly into the gel state, which is accompanied by a

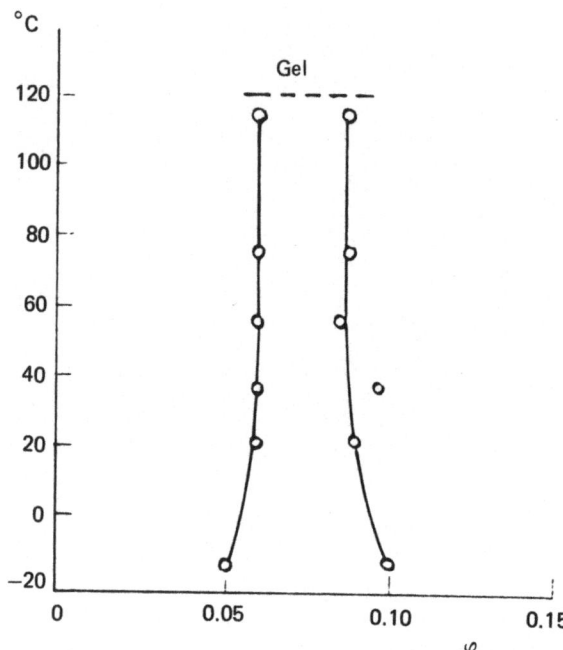

Figure 7.   Phase diagram of the system "poly(p-benzamide)-
            N,N-dimethylacetamide(+ LiCl)".

sharp increase in turbidity (light scattering). This
points to the process of crystallization of the polymer,
i.e., to the fact that the system is unstable with respect
to crystalline equilibrium.

Completion of the transition to the crystalline state
is also observed in fibers based on poly(p-phenyleneterephthalamide) during heat treatment, indicated, for example,
by the abrupt change in the properties of these fibers[5].

One more example of the formation of non-equilibrium
states in systems containing polymer with increased chain-rigidity is the well known case of spontaneous elongation
of films[6] and fibers[7] from cellulose acetate. As noted by
Flory[1], this is associated with the transition of the

polymer from the non-equilibrium amorphous state into the
ordered (liquid-crystalline) state. This transition of
the amorphous part of the regenerated cellulose fiber to
the ordered state evidently explains its spontaneous elongation on heating above the glass transition temperature[8].
The same phenomenon is observed for some other fibers made
from polymers with increased chain rigidity.

The general phase equilibrium diagram shown in Fig. 5
makes it possible to consider (due account being taken of
the phase transition kinetics) many practical cases of
separation of rigid chain polymers from solution when
obtaining fibers and films. This is particularly important
when studying the structural properties of the polymeric
materials obtained, and in controlling their properties.

REFERENCES

1. P. J. Flory, Proc. R. Soc., London, A234, 60, 73 (1956).

2. W. G. Miller, C. C. Wu, E. L. Wee, G. L. Santee,
    J. H. Rai, and K. G. Goebel, Pure Appl. Chem., 38, 37
    (1974).

3. V. G. Kulichikhin, et al., Vysokomol. Soedin. Ser. A.,
    18, 590 (1976).

4. S. P. Papkov, S. I. Bandurian and M. M. Iovleva, ibid.,
    Ser. B, 15, 370 (1973).

5. S. L. Kowlek, U.S. Pat. 3,671,542 (E. I. Du Pont
    de Nemours & Co., Inc.,1972).

6. T. G. Majury and H. L. Wellard, International Symposium
    on Macromolecular Chemistry, Torino, Italy, 1954.

7. B. A. Fomenko, L. P. Perepechkin, B. V. Vasil'ev, and
    N. I. Naimark, Vysokomol. Soedin., Ser. A, 11, 1971
    (1969)

8. A. T. Kalashnik and S. P. Papkov, ibid., Ser. B, <u>18</u>,
    455 (1976).

Editor's Note. Regrettably, Professor Papkov was
unable to attend the Symposium due to illness. I am sure
there would have been much valuable discussion of this
paper to report had he been there to present it. We are,
however, glad to print this paper, which reached our hands
shortly after the meeting, in these proceedings   (JRS).

LIQUID CRYSTAL POLYMERS. II. PREPARATION AND PROPERTIES OF

POLYESTERS EXHIBITING LIQUID CRYSTALLINE MELTS

F. E. McFarlane, V. A. Nicely, and T. G. Davis

Research Laboratories, Tennessee Eastman Company
Division of Eastman Kodak Company
Kingsport, Tennessee 37662

INTRODUCTION

The previous paper[1] in this series described the preparation and properties of highly aromatic polyesters that have turbid melts, have melt viscosities highly dependent upon composition and shear rate, and that give unusually anisotropic molded articles. Because these unusual properties are reminiscent of the behavior of nonpolymeric nematic liquid crystalline materials, further work has been done to synthesize and characterize polymers containing other moieties known to lead to liquid crystallinity in nonpolymeric materials. The copolyesters produced were derived by the acidolysis reaction previously described[1] from poly-(ethylene terephthalate) (PET) and a variety of dicarboxylic acids and acetylated difunctional phenols. Some of the copolymer compositions were varied to determine the limits of composition that give the turbid melts characteristic of liquid crystallinity, but which can be melted before decomposition. This paper describes the preparation and the physical and magnetic properties of these polymers.

EXPERIMENTAL

Materials

PET with an inherent viscosity of 0.55 to 0.65 was prepared from dimethyl terephthalate and ethylene glycol in a conventional manner with zinc acetate (65 ppm Zn) and antimony triacetate (250 ppm Sb) catalysts.[2]

The acetylated difunctional phenols were prepared by reaction of the phenols at 90 to 100°C with excess acetic anhydride containing a trace of sulfuric acid catalyst. The acetylated phenols were isolated and purified by distillation.

The following illustrates preparation of a typical liquid crystalline copolyester from reaction of PET with terephthalic acid, p-acetoxybenzoic acid, and hydroquinone diacetate. The following components were placed into a one-neck 500-ml round-bottom flask which was fitted with a stainless steel stirrer and a short head with a nitrogen inlet and condensate take-off.

> 38.4 g (0.2 moles) poly(ethylene terephthalate)
>   containing conventional zinc/antimony catalyst and
>   having an I.V. of 0.60
> 27.0 g (0.15 moles) p-acetoxybenzoic acid
> 24.9 g (0.15 moles) terephthalic acid
> 29.7 g (0.15 moles) hydroquinone diacetate

The condensate take-off was connected by ground glass fittings to a receiver cooled by dry ice and with provision for applying vacuum. After the flask was purged with nitrogen, it was immersed into a thermostatically controlled Wood's Metal bath at 150°C. The temperature was raised to 270°C over a period of approximately 30 min. A low viscosity clear melt was obtained which began evolving acetic acid when the temperature reached 260 to 270°C. After approximately 20 min at 270°C, the temperature was raised to 290°C, and the reaction continued for approximately 30 min. The pressure was reduced to 0.10-0.20 torr for 50 min. After cooling, the resulting copolyester had an I.V. of 0.75.

Methods

Inherent viscosities (I.V.) were measured at 25°C in 60/40 (v/v) phenol/tetrachloroethane at a concentration of 0.50 g/100 ml. NMR spectra were measured on trifluoroacetic acid solutions of the polyesters with a Brüker HX-90-E spectrometer.

The polymers were injection-molded into unheated molds in a 1-oz Watson-Stillman injection-molding machine to give $2^1/_2$- × $^3/_8$- × $^1/_{16}$-in. tensile bars for tensile measurements and 5- × $^1/_2$- × $^1/_8$-in. flexure bars for determination of flexural modulus, Izod impact strength, and heat-deflection temperature. Several compositions were also injection-molded in a 6-oz New Britain 175-TP reciprocating screw machine to give $8^{1}/_{2}$- × $^{3}/_{4}$- × $^{1}/_{8}$-in. tensile bars and in a Newbury HV1-25T reciprocating screw machine to give $6^{1}/_{2}$- × $^{3}/_{4}$- × $^{1}/_{8}$-in. tensile bars. ASTM procedures were used for measuring tensile strength and elongation to break (ASTM D1708), flexural modulus (ASTM D790), flexural strength (ASTM D790), Izod impact strength (ASTM D256 Method A), Rockwell hardness (ASTM D785 Method A), heat-deflection temperature (determined at 264 psi, ASTM D648), mold shrinkage (ASTM D955), and coefficient of linear thermal expansion (ASTM D696).

The viscosity of the polymers was measured on a Sieglaff-McKelvey rheometer by conventional techniques.

NMR spectra were collected on a 40-MHz spectrometer of conventional continuous wave design employing 35-Hz magnetic field modulation. The spectrometer was interfaced to a Digital Equipment Corporation PDP 8/S computer that was used to time average the spectra. The spectra were punched on paper tape and further processed on an IBM 1130 computer. For NMR analysis, the polymers, after being ground to pass a 3-mm screen, were placed in 11-mm Pyrex glass tubes which were dried overnight at 110°C, filled with nitrogen ($\leqq$ 1 atm), and then sealed by fusing the tube near the top.

Sample temperature was controlled by blowing gas at a constant rate over a resistance coil heated by a constant voltage supply and was measured by a copper constantan thermocouple at the sample with a Leeds & Northrup potentiometer (Cat. No. 8667). The temperature remained constant (± 1°) during the experiment. Because of the NMR probe design and heating technique, there was a temperature gradient of several degrees per centimeter over the samples. The spectra were checked for power saturation to assure that the relative areas were a meaningful measure of the number of protons.

Magnetic polarization of the polymer to give a net molecular orientation was produced by heating randomly oriented samples of the polymers in sample tubes (11-mm od) in the NMR probe in the 9400-gauss magnetic field. The polymers were cooled in the field by turning off the heater.

Macroscopic orientation was detected by measuring the NMR spectrum with the sample positioned as it was in the heated state (parallel orientation) or turned 90° (perpendicular orientation).

RESULTS AND DISCUSSION

Thermotropic liquid crystalline mesophases are usually recognized because the materials exhibiting them melt into a fluid which maintains the birefringent and light scattering properties characteristic of solids.[3-7] However, in these potentially inhomogeneous copolyesters, a demonstration of the characteristic three-dimensional translation, but one-dimensional rotation of nematic liquid crystal molecules would be a more definitive assignment of the type of phase present. We have obtained such evidence for selected polymers. A much larger set of copolyesters have been classified as liquid crystalline because they were turbid fluids that had easily induced shear orientation and which solidified after being stirred into characteristically tough, fibrous solids.

Effect of Composition on Liquid Crystallinity

It is well known that certain aromatic esters give nematic or smectic liquid crystals.[3-7] The temperatures of the solid-liquid crystalline and liquid crystalline-isotropic fluid transitions have been shown to be sensitive to the symmetry of the molecular structure, as influenced by side groups, and to the type of end groups on the molecule. Liquid crystallinity in polymeric molecules is affected by the additional parameter of composition just as liquid crystallinity of nonpolymeric materials is affected in solutions of the materials.

We have prepared copolymer compositions that have sufficient chain flexibility, or moiety dissymmetry, or a varied mixture of moieties to allow the polymer to melt at or below 300°C and yet exhibit the characteristic melt turbidity and tough, fibrous behavior of solidified materials obtained from the flow-oriented nematic mesophase. Similar copolyesters with this behavior were prepared to illustrate the molecular characteristics important in determining when a mesophase will be observed.

The copolyester of p-oxybenzoyl (I) and ethylene
terephthalate (II) moieties were synthesized in 10 mol %

(I)   (II)

increments from 0 to 80 mol % I. Compositions with 30 mol %
or less of I melted as ordinary copolymers to give clear,
isotropic fluid phases. However, 40, 50, 60, 70, or 80 mol %
I gave a copolyester that was turbid in the melt and became a
tough, fibrous solid when cooled shortly after shearing of the
melt. If the melt was allowed to stand without stirring for a
period and then cooled, however, the resulting solid was brittle.
Above 80 mol % I, a high melting, low molecular weight polymer
was formed by conventional melt-phase polymerization techniques,
although solid-phase polymerization resulted in a useful, high
molecular weight polymer.[1]

Trifluoroacetic acid solutions of copolyesters containing
varying amounts of the moieties (I) and (II) gave resolved
13-carbon NMR resonances for the three distinct types of
carbonyl linkages in the copolymer. The measured number of
dyads of I with itself compared with the number expected in a
random copolymer of I and II, as shown in Figure 1, indicated
that the described preparation leads to a random copolymer.
Therefore, random copolymers of I and II can have liquid
crystalline mesophases as judged by optical and orientation
behavior. The polymers yielding turbid melts do not undergo
a liquid crystalline to isotropic transition before decomposition. These observations are consistent with behavior
of previously reported nonpolymeric liquid crystals containing
I.[8] Such liquid crystals are characterized by relatively
high melting points and broad nematic phase temperature ranges
that increase with increasing molecular weight.[8] The
asymmetry of I and the chain flexibility induced by II tend
to lower the copolymer's melting point; but the planar, para-
orientation of I leads to liquid crystallinity in the high
molecular weight copolymer.

A similar series of copolyesters have been made from
moieties II and III.

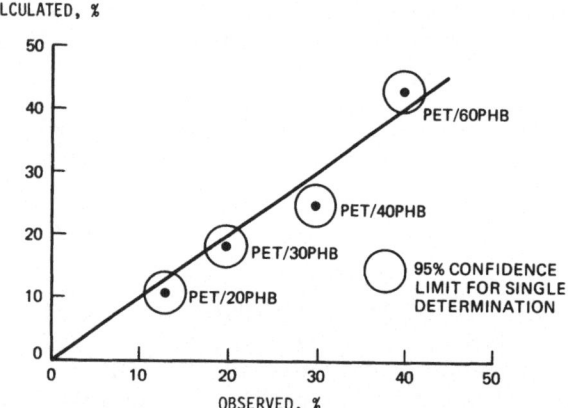

Fig. 1.   Percent of the p-oxybenzoyl moieties bonded to another p-oxybenzoyl moiety as measured by 13-carbon NMR and predicted by statistical random distribution for a copolyester made from PET and different percentages of p-hydroxybenzoic acid.

(III)

At concentrations of III from ca. 40 mol % to ca. 70 mol %, the polymer gave turbid melts and fibrous, tough solids, but was a nonmelting solid at 290°C above ca. 80 mol % III.   Similar behavior was observed for copolyesters of II and IV.

(IV)

These copolyesters had melting points in the desirable range because of the chain flexibility introduced by II and the asymetry of III or IV.

One further composition series was the copolyester of II
and V which also gave turbid melts above ca. 15 mol % V.

(V)

However, because of the symmetry of V, high melting copolyesters
resulted above ca. 35 mole % V.

Tables 1, 2, and 3 show the properties of many compositions which have characteristics that we associate with
nematic, liquid crystalline melts. Several of the compositions
were obtained by modification of the copolyester of I and II
by replacement of a fraction of I or II by other planar,
para-substituted moieties expected to yield liquid crystalline
polymers. Although the examples represent only a small
fraction of the many compositions prepared, they clearly
illustrate the generality of the phenomenon under study. As
the number of dicarboxylate and diol moieties increases, the
number of distinct repeat units, i.e., pairs connected in
the sense of units II, III, or IV, grows as m*n, where m is
the number of dicarboxylates and n is the number of diols.
Therefore, several of the copolymers have five distinct
repeat units potentially present. We have synthesized
liquid crystalline polymers having as many as eight different
repeat units, so complexity cf composition does not appear
to severely limit the polymers which can exhibit the behavior.

Table 4 shows some typical compositions which do not
exhibit a liquid crystalline mesophase. For example, replacing
I by VI or II by VII decreases the linearity of the polymer

(VI)   (VII)

TABLE 1. Physical Properties of Typical Liquid Crystalline Polyesters[a]

Polymer Structure	Inherent Viscosity (I.V.)	Tensile Break, 10^3 psi	Break Elong, %	Flexural Modulus, 10^3 psi	Izod Impact V-Notch	Izod Impact Un-notched	Heat Distortion (264 psi), °C
	0.62	7.2	243	3.3	0.3	9.5	66
	0.64	34.3	9	19	4.4	24.0	65
	0.96	31.8	22	12	4.7	12.6	74
	0.83	29.5	24	16	3.9	15.8	87
	0.62	24.8	16	12	1.6	4.6	70
	0.68	25.2	21	13	3.7	14.0	105
	1.04	37.3	31	15	5.0	18.8	102

[a]Molding temperatures vary between 260 and 290°C depending on polymer composition.

TABLE 2. Physical Properties of Typical Liquid Crystalline Polymers[a]

Polymer Structure	Inherent Viscosity (I.V.)	Tensile Break, 10³ psi	Break Elong, %	Flexural Modulus, 10⁵ psi	Izod Impact V-Notch	Izod Impact Un-notched	Heat Distortion (264 psi), °C
(structure)	0.87	20.2	16	14.0	2.0	9.3	99
(structure)	0.88	37.9	17	17.6	3.0	11.2	85
(structure)	1.10	39.0	11	16.5	5.4	11.4	115
(structure)	0.82	39.5	18	17.4	9.17	11.1	126
(structure)	0.78	33.7	12	21.9	2.9	7.2	93
(structure)	0.83	33.9	13	18.2	3.1	7.7	98

[a]Molding temperatures vary between 260 and 290°C depending on polymer composition.

TABLE 3. Effect of a Nonlinear Structural Unit on Polymer Physical Properties

Polymer Structure	Molding[a] Temp,°C	Inherent Viscosity (I.V.)	Tensile Break, 10^3 psi	Break Elong, %	Flexural Modulus, 10^3 psi	V-Notch Izod Impact
1.	260	0.65	34.3	9	17.8	4.4
2.	260	0.62	31.0	22	14.3	1.8
3.	260	0.70	21.8	22	10.6	0.3
4.	290	0.51	8.4	61	3.0	0.8
5.	290	0.68	25.2	21	13.0	3.7
6.	290	0.69	20.8	27	14.0	0.16

[a]Molded in a Watson-Stillman molding machine.

TABLE 4. Examples of Polymers Giving Clear, Nonturbid Isotropic Melts

chain sufficient to completely remove the liquid crystalline
mesophase behavior. Also, the introduction of substantial
amounts of the nonplanar VIII also results in polymers having
no liquid crystalline mesophase.

$$-O-\bigcirc-\overset{\overset{\displaystyle CH_3}{|}}{\underset{\underset{\displaystyle CH_3}{|}}{C}}-\bigcirc-O-$$

(VIII)

Wide-Line NMR Evidence of Liquid Crystallinity

    The NMR spectrum line shape of solid or highly viscous
liquid materials is due primarily to the nuclear-nuclear di-
pole coupling of the nuclei averaged over their motional
and orientational distribution functions.[9] The spectrum of
a material which has no molecular rotation gives a broad
line with the detailed shape dependent upon the molecule, its
orientation distribution, and the sample orientation in the
magnetic field. A material in which the molecules
rotate very rapidly about one axis and very slowly about
another axis gives a spectrum exhibiting extreme line narrowing
with respect to the rapid rotation. Therefore, a solid-like
spectrum occurs because of the residual dipole-dipole coupling,
but the spectrum is unique because of the partially averaged
dipole-dipole coupling. A spectrum from molecules having
such anisotropic rotation is capable of showing magnetic
anistropy, just like an ordinary solid, if a nonspherical
macroscopic orientation distribution of molecular axes exists
in the sample.

    The molecules in a nematic mesophase tend to align along
an external magnetic field.[10] Thus, the induction of a net
molecular alignment by a magnetic field would be indicative
of a nematic mesophase structure. However, because small
anisotropic crystals suspended in a liquid might also be
oriented by an external field, magnetic field-induced
molecular orientation would not be conclusive proof of a
nematic mesophase structure. The molecules in a nematic
mesophase rotate rapidly about only one axis and translate
rapidly in three dimensions.[10,11] These unusual molecular
motional characteristics would be most easily confirmed by
their effects on the wide-line NMR spectrum because of the

partial averaging of the magnetic dipole-dipole coupling caused by this anisotropic motion.[10],[11] Furthermore, line shapes expected for these motional characteristics are qualitatively different from those expected for crystals suspended in a liquid. Thus, magnetic polarization of the mesophase and observation of the unusual partial averaging of the magnetic dipole-dipole interactions, together, would provide strong evidence that the melt was a nematic mesophase rather than crystals in a liquid or an emulsion of immiscible liquids.

A typical temperature sequence between room temperature and 300°C for the NMR spectra of the liquid crystalline polymers is as follows: At room temperature a single broad featureless line, similar to those shown in Figure 2, is observed. As the temperature is increased, a narrow line appears and is centered on the broad line which becomes more narrow. Further increases in temperature cause an increasing fraction of the signal to appear in the narrow line, but the broad line maintains its shape.

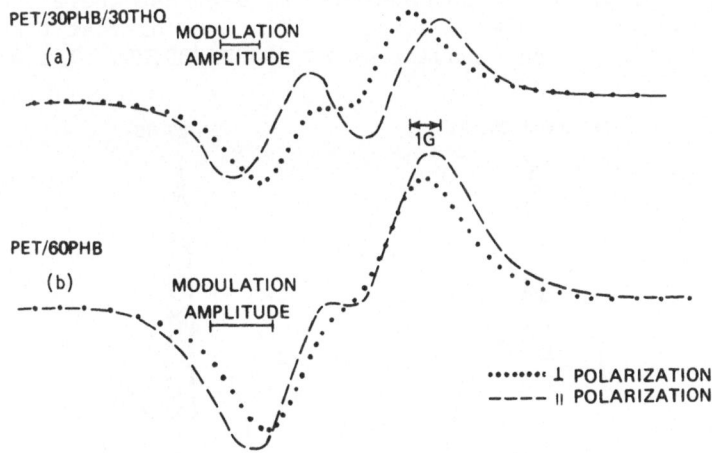

Fig. 2. First-derivative wide-line NMR spectra of magnetically polarized PET/30PHB/30THQ (the copolyester made from PET, 30 mol % p-hydroxybenzoic acid, and 30 mol % of the repeat unit derived from terephthalic acid and hydroquinone) and PET/60PHB (the copolyester made from PET and 60 mol % p-hydroxybenzoic acid) at room temperature for parallel (ǁ) and perpendicular (⊥) relations of the polarization and observation magnetic fields.

Figure 3 shows typical NMR curves at a temperature where both narrow and broad lines were observed. The displayed width of the narrow line is due to the magnetic field modulation amplitude, but its natural width is much less. The narrow lines were separated by the dotted line illustrated in Figure 3 and integrated by computer procedures to find their areas relative to the total area. These relative areas are plotted as a function of temperature in Figure 4 for several polymers. The spectra in Figure 3 have two peaks marked "a" and two shoulders marked "b". The distance between the two "a" peaks and between the two "b" peaks is plotted in Figure 5 for PET/60PHB and PET/30PHB/30THQ.

All of the polymers began to show a narrow component at nearly the same temperature according to Figure 4. In PET/30PHB, the fraction of protons in the narrow line increased rapidly with increasing temperature so that by 225°C there was no broad line apparent. At 225°C, PET/30PHB was a clear melt. Both these observations are typical of a polymer melting to give a normal liquid. Table 5 shows that the viscosity of PET/30PHB continues to decrease above the melting temperature of about 220°C. Now contrast that behavior with the behavior of PET/30PHB/30THQ. Its spectrum

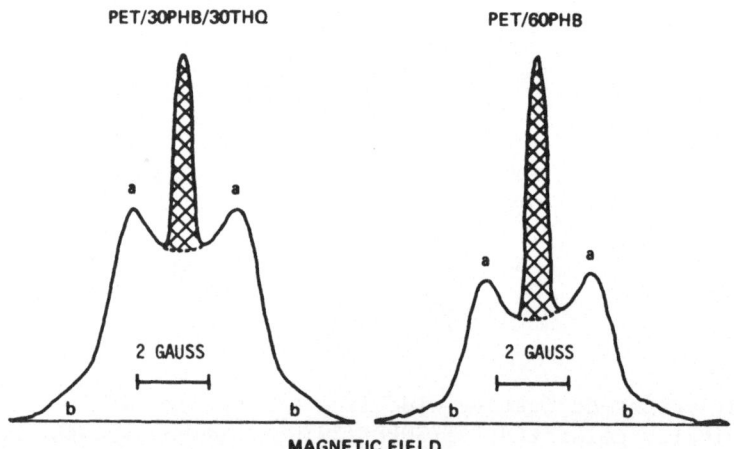

Fig. 3. Typical melt phase spectra for two polymers at 240°C, where a and b indicate distinctive features of the broad line. The dashed line indicates the separation of narrow and broad line.

had a much more gradual increase of narrow component with increasing temperature so that at 350°C only 15% of its protons exhibited a sufficiently isotropic rotation to create a narrow line. Yet, data in Table 5 show that by 270°C, its viscosity was as low as that of the PET/30PHB at the same temperature and lower than that of the PET/30PHB as an isotropic melt at 230°C. Furthermore, as the viscosity of PET/30PHB/30THQ dropped rapidly (Table 5), the fraction of protons in the narrow component, that is, the amount of isotropic liquid, changed very little (Figure 4). PET/60PHB is very similar in both its NMR and viscosity behavior to PET/30PHB/30THQ. Therefore, this evidence indicates that the isotropically rotating moieties are probably not responsible for the fluidity of PET/60PHB or PET/30PHB/30THQ. The other polymers shown have behavior intermediate between the extremes just discussed.

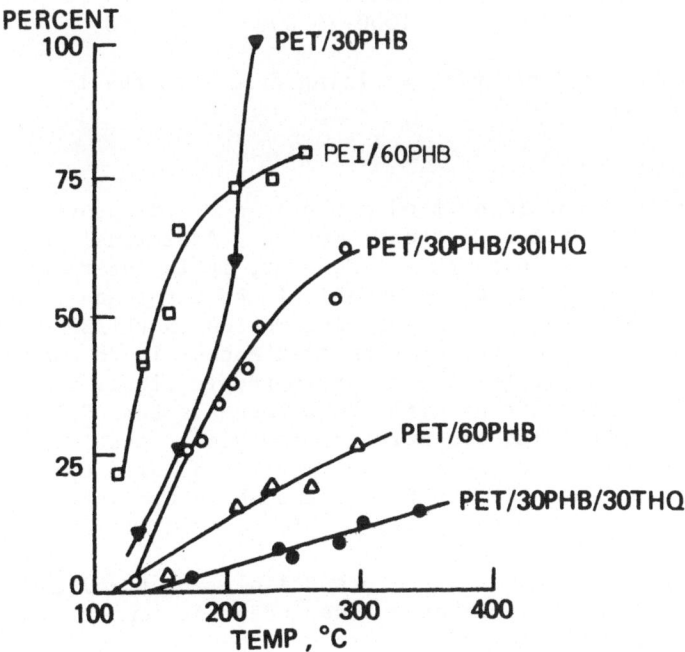

Fig. 4. Percent of total area in the narrow line vs. temperature. PEI is poly(ethylene isophthalate).

TABLE 5

Approximate Polymer Melt Viscosities at Several Temperatures

Polymer	Temp, °C	Shear Rate Range, sec^{-1}	Viscosity Range, poise
PET/60PHB	230	3–62	22,809–5800
	270	520–1500	30–71
	300	1000–10,000	25–
PET/30PHB	230	10–100	3900–4700
	270	383–1000	65–104
	300	1000–5000	34–47
PET/30PHB/- 30THQ	230	2–39	345,700–30,000
	270	1200–3100	44
	300	1000–10,000	10
PET/30PHB/- 30IHQ[a]	230	20–136	6200–2600
	270	100–2400	63–76
	300	1000–10,000	50–

[a]IHQ is the repeat unit resulting from isophthalic acid and hydroquinone.

We have not identified the origin of the protons responsible for the isotropic motion. Additional work will be necessary to learn this. However, it is interesting to note that at 350°C, not even all of the methylene protons can execute a spherical averaging motion in PET/30PHB/30THQ because there are more of them than appear in the narrow component. Thus, the amount of isotropic liquid observed in the melt is consistent with the hypothesis that PET/60PHB and PET/30PHB/30THQ are nematic mesophases, that PET/30PHB is a normal semicrystalline polymer which melts to an isotropic liquid, and that the other polymers shown are intermediate cases.

The line shapes of the broad line were found to be very similar and remarkably independent of temperature for the polymers showing small narrow components. Figure 3 illustrates the observation. The separation of the peaks marked "a" and "b" is shown in Figure 5 for a range of temper-

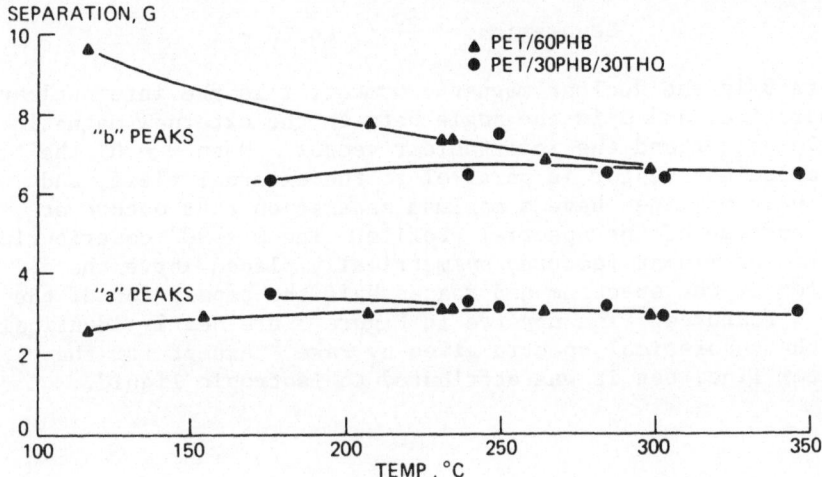

Fig. 5. Separation of peaks in melt-phase spectra vs. temperature.

atures. These shapes are different from that of the solid polymer (Figure 2), but once this narrow shape is achieved further increases in temperature do not change it. Even magnetic polarization changes only the relative area in these peaks as shown in Figure 6, but not their separation. The additional empirical observation from Figure 5 is that the "b" peaks are separated by nearly twice the separation of the "a" peaks. This type spectrum is not due to chemical shifts nor spin-spin coupling as in high resolution NMR because the splitting is much too large. Actually, the shape of the pattern and the constancy over a range of temperature is very similar to that observed for nematic liquid crystals.[11-15]

Now we shall consider a semiquantitative theory of the shape of the broad component of the melt spectrum. Since over 70% of the protons are on para-substituted aromatic rings in PET/60PHB and PET/30PHB/30THQ, these protons will contribute most of the spectrum. As a first approximation, the neighboring protons on each ring can be considered to be an isolated proton pair. Pake[16] has discussed the spectrum for an isolated pair of protons. Each pair contributes two lines centered at the same position and separated by

$$\text{Separation} = \frac{3u}{2r^3}(3\cos^2\theta - 1)$$

where u is the nuclear magnetic moment, r is the internuclear separation, and Θ is the angle between the external magnetic field vector and the internuclear vector. When Θ = 0° the internuclear vector is parallel to the external field, and the pair of lines have a maximum separation that occurs at the extreme of the spectral profile. The Θ = 90° contribution is two prominent features symmetrically placed above the center of the spectrum and spaced half the separation of the Θ = 0 features. The spectra in Figure 3 are nearly identical to the theoretical spectra given by Pake[16] except for the center line, but it was attributed to isotropic liquid.

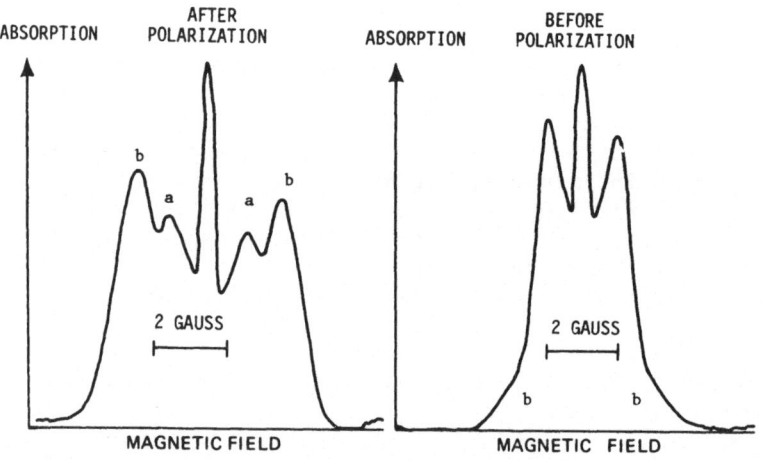

Fig. 6. Effect of magnetic field polarization on melt spectrum of PET/30PHB/30THQ. Polarized at 284°C in 9400-gauss field and spectrum obtained at 239°C.

Eichhoff and Zackmann[17] have tabulated the proton-proton distances in poly(ethylene terephthalate). Because the nearest neighbor protons on the terephthalate ring are 2.21 Å apart, while there is 4.74 Å between those on opposite sides of the ring, and because the splitting is inversely proportional to this distance cubed, only the neighboring pairs have a strong interaction. Thus, if the molecule is rotating about one axis and translating rapidly in three dimensions to average intermolecular interactions, the spectrum should look very similar to that of isolated proton pairs. In fact, a detailed calculation of the powder pattern for terephthalate rings rotating about their symmetry axis does give a very similar result to the one discussed.[17]

From the model and data we can make several predictions about the observed spectra. First, the expected separation of the ⊥ peaks (labeled a) in Figure 3 is predicted to be 4 gauss and the separation of the ‖ peaks (labeled b) is predicted to be 8 gauss for the proton-proton separation of 2.21 Å of the terephthalate ring. According to the data shown in Figure 5, that is very near the observed separation. Second, we can predict from the model that the shoulders on the outer edges of the spectra are due to contributions of molecules parallel to the external field. When magnetic orientation occurs, the number of molecules with their long axis parallel to the external field increases, so these shoulders should become more prominent. Figure 6 shows that this does occur. The observed spectrum has the proper qualitative shape as predicted by the model. Thus, all the data are consistent with the hypothesis that the melts of PET/60PHB and PET/30PHB/30THQ are primarily composed of molecules rotating about one axis and translating in three dimensions.

Finally, nematic liquid crystals can have a macroscopic molecular orientation induced by external electric or magnetic fields. Because of the high viscosity at low shear for the polymer studied, high temperatures and long times were required to induce measurable molecular orientation. However, as illustrated in Figures 2 and 6 several of the polymers could be oriented in a magnetic field. The magnetic anisotropy was used to monitor the induced orientation. Figure 2 shows some examples of the spectra of polymers that had been solidified while polarized. In agreement with the

qualitative model previously proposed, the parallel spectra
were the broadest because the oriented polymers have a
greater than average number of molecules parallel to the
polarization direction. The line shape is different from
that in the melt-phase spectra because these spectra of
solids have intermolecular dipole-dipole coupling as well as
the intramolecular dipole-dipole coupling previously discussed.
Data are given in Table 6 for three of the polymers polarized
by a magnetic field. This behavior is indicative of nematic
liquid crystal structure in the melt.

In summary, the four types of wide-line NMR data that
show these polymers behave as nematic liquid crystals are:
(1) amount of narrow line, or the fraction of protons
capable of isotropic rotation in the liquid phase, (2) lack
of line shape change over a broad melt temperature range,
(3) magnetic field-induced molecular orientation, and
(4) ideal line shape for a liquid of axially rotating,
three-dimensionally translating molecules.

TABLE 6

Results of Magnetic Orientation

| Polymer | Polarization | | ΔH_{pp}, G^{a} | |
	Temp, °C	Time, Min	$\parallel^{b}$	$\perp^{b}$
PET/60PHB	300	60	5.83	5.04
PET/30PHB/30THQ	320	40	5.16	4.65
PET/30PHB/30IHQ[c]	320	60	5.55	5.17

[a] ΔH_{pp} is the peak-to-peak line width (in gauss) of the first
derivative NMR spectrum at room temperature uncorrected for
modulation amplitude.

[b] $\parallel$, $\perp$ refer to the direction of the magnetic field during the
measurement relative to its direction during the high
temperature polarization.

[c] IHQ is the repeat unit resulting from isophthalic acid
and hydroquinone.

Viscosity of Melts

The dependence of the melt viscosity on composition and shear rate is also a characteristic of liquid crystalline polymers. Figure 7, from Jackson and Kuhfuss,[1] shows the melt viscosity of a series of PET polymers modified with varying amounts of PHB, but of comparable molecular weights. The melt viscosity increases with increasing PHB content up to 30 mol % PHB. This is due to the replacement of relatively flexible ethylene terephthalate units with relatively inflexible p-oxybenzoyl units. However, the decrease in melt viscosity that occurs at PHB contents > 30 mol % PHB is not anticipated according to the usual correlations of molecular structure with melt viscosity.

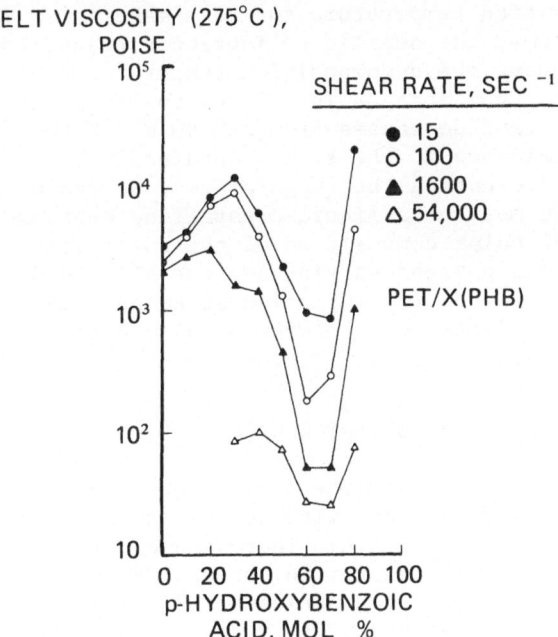

Fig. 7. Melt viscosity of PET modified with p-hydroxybenzoic acid. © John Wiley & Sons, Inc., 1976. Reproduced with permission.

We noted during polymer preparation that clear, transparent melts were obtained until the mole percent PHB in the copolymers exceeded 30 mol %. Above 30 mol % of PHB, turbid melts were obtained with the effect most pronounced in polymers containing 60 to 70 mol % PHB. Until the mole percent PHB in PET exceeds 30 mol %, a sufficient concentration of p-oxybenzoyl units is not present to form liquid crystals and a clear, isotropic polymer melt results. Beyond 30 mol % PHB, the structural moieties capable of forming liquid crystals are present in sufficient concentration to begin imparting liquid crystalline characteristics to the polymers.

An increase in melt viscosity on going from a nematic mesophase to an isotropic liquid has been reported by Porter and Johnson for p-azoxyanisole, a well characterized nematic liquid crystalline compound.[18] They observed a marked increase in the viscosity of the flow-oriented nematic phase as the temperature was increased through the 135°C nematic to isotropic transition temperature for this material. In our polymeric systems, the nematic to isotropic transition temperatures are above the decomposition temperatures of the polymers. However, the increase in melt viscosity shown in Figure 7 as PHB content decreases from ca. 60 mol % to ca. 30 mol % is analogous to the result obtained with p-azoxyanisole by increasing the temperature. We can obtain either an isotropic melt or a liquid crystalline melt by controlling the molecular composition of the polyester. Polymeric systems can possess varying degrees of liquid crystallinity depending on their particular composition. This is an important difference between nonpolymeric and polymeric liquid crystals.

Quenched-Polymer Density

The onset of liquid crystallinity in a copolymer composition series is coincident with an increased degree of order in the melt. Therefore, this increased order might be observed as an increase in density in pressed films quenched from the melt. The density of a series of PET polymers modified with varying amounts of PHB is shown in Figure 8. The samples were prepared by pressing films at 295°C, and then quenching them in ice water. We know that this technique effectively quenches PET and provides amorphous films. Thus, any density changes must reflect changes resulting from the PHB content of the polymer. The data show that the

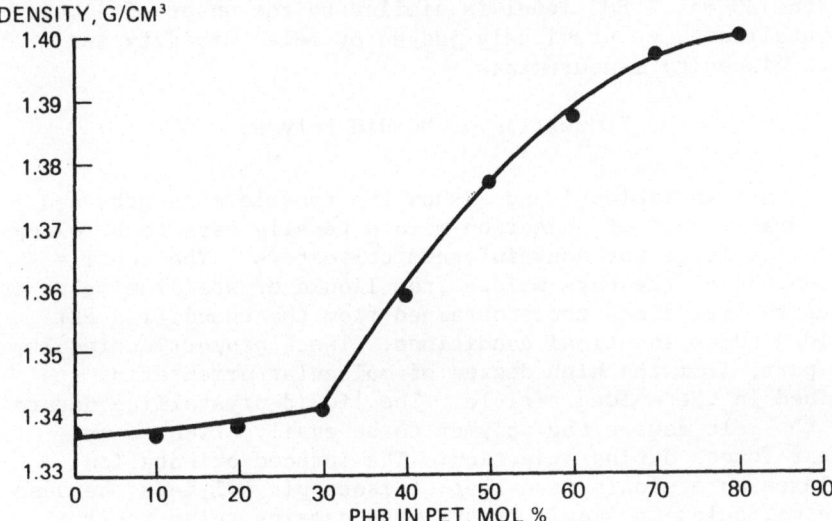

Fig. 8   Quenched-polymer density of PET/PHB copolymers.

density of PET/PHB copolyesters maintains a relatively low, constant value until the PHB content reaches 30 mol %.   From 30 mol % to ca. 80 mol % PHB, the density of the film increases rapidly.   This increase in density occurs despite the fact that X-ray diffraction showed no appreciable amount of crystalline material present in the films until the PHB content of the copolymer approached 80 mol %.

Onsager[19] first discovered that a system of long rigid rods showed a transition from an isotropic phase to a denser aniostropic phase.   Lattice models were later formulated by Flory[20] and Dimarizio[21] which gave results in semiquantitative agreement with Onsager's theory.   The increase in density that occurs as the mole percent PHB increases from 30 to ca. 80 is the result expected for a polymer changing from an isotropic fluid to an anisotropic fluid, the degree of order in the melt increasing with increasing PHB content of the polymer.   Although the densities of these quenched solids do not show that the densities of the melts vary in the same manner, they do indicate that a high fraction of PHB can lead to a more dense, noncrystalline solid.   The change in density with changing PHB content for PET/PHB copolyesters

at the 30 mol % PHB level is similar to the onset of liquid
crystallinity as previously judged by melt turbidity and
melt viscosity measurements.

Properties of Molded Polymer

Data in Tables 1 and 2 show the tensile strengths and
flexural moduli of injection-molded tensile bars to be exceptionally large for nonreinforced polyesters. The tensile
strengths of the bars molded from liquid crystalline melts are
four to five times those obtained from the unmodified PET
molded under identical conditions. These properties result,
in part, from the high degree of molecular orientation retained in the molded article. The liquid crystalline nature
of the melt causes the polymer to be easily oriented by
shear forces during injection. The induced orientation
relaxes more slowly than that of isotropic polymers, because
the molecules in liquid crystalline domains relax their
molecular orientation cooperatively so that the effective
mass being relaxed by thermal energy is much greater than it
is in isotropic polymers. Therefore, the observed average
molecular orientation in a molded article may depend upon
the polymer's melt temperature, the mold's temperature and
mass, the article's shape and dimensions, and the polymer's
flow pattern and injection rate into the mold.

Table 7 shows the direction-dependent properties of
injection-molded $^1/_8$-in. plaques that were gated along one
edge of the mold. The anisotropic properties are due
to the quenched molecular orientation in the plaque; the
increased property values were obtained from the bars cut
from the plaque in the direction parallel to the polymer flow.
This explanation is further substantiated by measured proton
magnetic resonance anisotropy.[22]

The physical properties obtained for these liquid crystal
polymers are affected by both the inherent viscosity (mol wt)
of the polymer and the temperature of the melt during injection molding as Table 8 shows. As the molding temperature
for PET/30PHB/30THQ increased from 240 to 280°C and the inherent viscosity of the polymer increased from 0.50 to 0.94, the
tensile strength and flexural modulus increased by approximately a factor of 3, with very substantial increases in Izod impact strenght also occurring. It is interesting that an increase in physical properties is accompanied by a decrease in
mold shrinkage as shown in Table 8.

TABLE 7

Anisotropic Physical Properties of PET/60PHB
and PET/30PHB/30THQ Polymers

Property	PET/60PHB[a]		PET/30PHB/30THQ[b]	
	Across Flow	With Flow	Across Flow	With Flow
Flexural modulus, 10^5 psi	2.3	17.1	2.4	14.4
Flexural strength, 10^3 psi	4.9	15.9	6.3	16.8
Izod impact (V-notch), ft-lb/in.	0.6	6.1	0.9	1.7
Izod impact (unnotched), ft-lb/in.	2.9	9.9	2.8	6.3
Tensile strength, 10^3 psi	4.2	15.5	5.4	14.8
Break elongation, %	10	8	3	3
Coef thermal expansion in./in./°C	4.5×10^{-5}	0	6.0×10^{-5}	0
Shrinkage, %	0.3	0	0.5	0
I.V. after molding	0.61	--	0.78	--

[a] $4^1/_2$- × $4^1/_2$- × $^1/_8$-in. plaques, gated along one edge and
molded in a 6-oz New Britian 175-TP machine (cylinder
temperature 260°C, mold temperature 23°C), were cut into
0.5-in. wide specimens; some were cut from the plaques
across the direction of polymer flow and some were cut
parallel to the direction of the flow. These specimens
were milled into the standard tensile bar shaft for tensile
measurements.

[b] Cylinder temperature 290°C; otherwise as in (a).

TABLE 8

Effect of Molding Temperature and Polymer I.V. on the Physical Properties of PET/30PHB/30THQ[a]

Melt Temp, °C	Polymer I.V.	Tensile Break, 10^3 psi	Break Elong, %	Flex. Modulus, 10^5 psi	Izod Impact V-notch	Izod Impact Un-notched	Heat Distortion (264 psi), °C	Mold Shrinkage, %
240	0.50	10.6	9	6.5	0.18	2.24	77	0.48
280	0.50	14.5	9	12.5	0.64	3.92	88	0.10
240	0.69	13.3	10	8.4	0.52	3.94	139	0.16
280	0.69	23.3	15	12.3	2.24	8.59	144	0.04
240	0.94	26.2	19	14.8	2.46	15.0	104	0.06
280	0.94	32.3	20	16.5	4.48	16.2	120	0

[a]Molded in Watson-Stillman molding machine, 23°C mold.

The polymers listed in Table 3 reveal the effect which structural units that do not form liquid crystals can have on the physical properties of molded bars. The first entry is included for comparison purposes only. The data given for entries two and three show a significant decrease in tensile break strength, flexural modulus, and Izod impact strength (V-notch) when part of the terephthalate is replaced with isophthalate moiety. The molecular weights of entries one through three are comparable and the samples were molded under identical conditions. It is also instructive to compare the data obtained on polymers five and six. Again we see that replacement of part of the terephthalate moiety with isophthalate results in a decrease in tensile strength and Izod impact strength, although the flexural modulus remained unchanged for this sample. The decrease in physical properties is believed to result from a decrease in the liquid crystalline nature of the polymers resulting from the presence of the nonliquid crystal forming isophthalate unit. The liquid crystalline nature of the polymer melt is important in controlling the amount of molecular orientation achieved during the injection molding operation. The orientation thus achieved has large effects on the physical properties of molded articles.

Whereas, a decrease in liquid crystallinity resulted from replacement of terephthalate with isophthalate, comparison of entries four and five reveals that replacement of hydroquinone with bisphenol A yields a completely nonliquid crystalline polymer. The physical properties of the polymer containing bisphenol A (entry four) are those typically obtained for a largely aromatic, amorphous polyester. This polymer gives a transparent, high-viscosity melt in contrast to the opaque, relatively lower viscosity melts of the liquid crystalline polymers. The bisphenol A unit, with a tetrahedral carbon linking the aromatic rings, is a nonplanar molecule with a 109° bend in the middle. Nonplanar structures of this type do not normally form liquid crystals. However, it is not obvious why polymer containing 30 mol % bisphenol A yield completely isotropic melts while polymers containing planar, nonlinear isophthalate units retain significant liquid crystalline character. The wideline NMR data previously given confirm that isophthalate-containing polymers of the type shown in Table 3 retain significant liquid crystalline character.

Figure 9 shows the effect of polymer composition on the physical properties of a series of PET copolymers containing p-hydroxybenzoic acid and hydroquinone. The tensile strength, flexural modulus, and Izod impact strength all have maximum values for polymers containing about 50 mol % PET and 50 mol % PHB/THQ combined. The increase in physical properties is believed to reflect the increase in liquid crystallinity of the polymer melts as the combined PHB/THQ components approach 50 mol %. The lower physical properties values obtained at the 70 to 80 mol % PHB/THQ levels may result from incomplete melting of the polymers at the 280°C melt temperature.

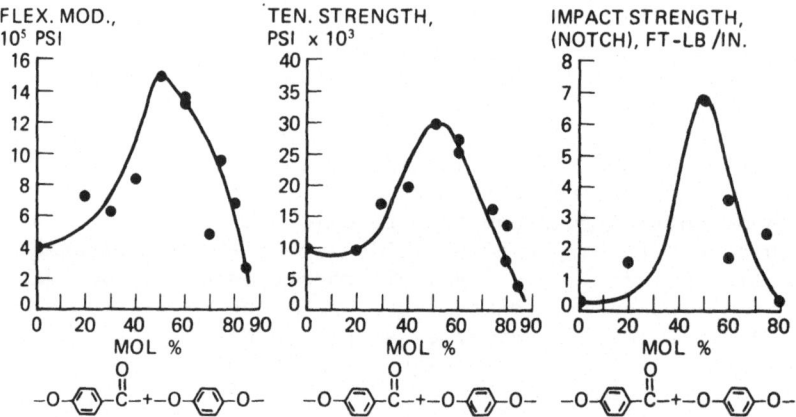

Fig. 9. Physical properties of PET/PHB/THQ copolymers. Specimens molded in a Watson–Stillman molding machine at 280°C melt temp and 23°C mold. temp; mol % PHB same as mol % THQ.

CONCLUSION

A wide variety of high molecular weight copolyesters exhibiting liquid crystalline melts have been synthesized. The preparations involve the acidolysis of PET with appropiate dicarboxylic acid and acetylated diphenols. The effect of molecular structure and composition on polymer melt viscosity, physical properties, and the wide-line NMR characteristics of the melt have been explored. The physical properties of injection-molded tensile bars are exceptionally high for non-reinforced polyesters. The mechanical properties are highly anisotropic and depend upon the polymer molecular weight, degree of liquid crystallinity, melt temperature during injection molding, and upon the mold's shape, dimensions, temperature, and melt flow pattern. These effects are shown to result from the liquid crystalline nature of the melt which allows production of molded articles possessing a high degree of uniaxial molecular orientation. The data are sufficient to warrant classification of these materials as liquid crystalline polyesters which represent a new class of polymers with novel and interesting properties.

REFERENCES

1. W. J. Jackson and H. F. Kuhfuss, J. Polym. Sci., 14, 2043 (1976).

2. J. G. Smith, C. J. Kibler, and B. J. Sublett, J. Polym. Sci., Part A-1, 4, 1851-1859 (1966).

3. J. A. Castellano and G. H. Brown, Chem. Technol., 47-52, 229-235 (1973).

4. G. W. Gray, "Molecular Structure and Properties of Liquid Crystals," Academic Press, New York, N.Y., 1962.

5. G. H. Brown and W. G. Shaw, Chem. Rev., 57, 1049 (1957).

6. A. Saupe, Angew. Chem. Int. Ed. Engl., 7, 97 (1968).

7. S. E. B. Petrie, H. K. Bücher, R. T. Klingbiel, and P. I. Rose, Eastman Org. Chem. Bull., 45 (No. 2) (1973).

8. D. Volander, Z. Physiol. Chem., 105, 211-254 (1923).

9. C. P. Slichter, "Principles of Magnetic Resonance," Harper and Row, New York, 1963, pp 45-63.

10. G. H. Brown, J. W. Doane, and V. D. Neff, "A Review of the Structure and Properties of Liquid Crystals," CRC Press, Cleveland, Ohio, pp 49-56.

11. N. Bravo, J. W. Doane, S. L. Arora, and J. L. Fergason J. Chem. Phys., 50, 1398 (1969).

12. R. D. Spence, H. A. Moses, and P. L. Jain, J. Chem. Phys., 18, 162 (1953).

13. R. D. Spence, H. S. Gutowsky, and C. H. Holm, J. Chem. Phys., 21, 1891 (1953).

14. P. L. Jain, J. C. Lee, and R. D. Spence, J. Chem. Phys., 23, 878 (1955).

15. T. J. Flautt and K. D. Lawson, in "Ordered Fluids and Liquid Crystals," R. S. Porter and J. F. Johnson, Symposium Chairmen, Advances in Chemistry, Series 63, American Chemical Society, Washington, D.C., 1967.

16. G. E. Pake, J. Chem. Phys., 16, 327 (1948).

17. U. Eichhoff and H. G. Zachmann, Kolloid-Z. Z. Polym., 241, 928 (1970).

18. R. S. Porter and J. F. Johnson, J. Appl. Phys., 34 (1), 51 (1953).

19. L. Onsager, Proc. Rochester Acad. Sci, 51, 627 (1949).

20. P. J. Flory, Proc. Roy. Soc. (London), A234, 73 (1956).

21. E. A. Dimarzio, J. Chem. Phys., 35, 658 (1961).

22. V. A. Nicely, unpublished data.

DISCUSSION

J. R. SCHAEFGEN - DU PONT. How do you account for the difference in properties with variation in thickness of your moldings? Have you sectioned them to see if there is a difference in orientation as you go through the samples? F. E. MCFARLANE - TENNESSEE EASTMAN CO. Yes, John, there certainly is. In a fractured sample visual observations show different flow patterns on the surface as compared to the interior of the sample. The orientation is highest on the surface of a tensile bar with significantly less orientation in the interior. Therefore, if you make thinner tensile bars the properties go up by a factor of two or more. So when you talk about physical properties, one must keep in mind that the manner in which bars are molded is very critical. Indeed, the real challenge now is for the engineering to catch up with the chemistry and to decide the best applications for these polymers. But certainly the properties of the same polymer can be changed from nominal to super depending on how you process it.

R. NATARAJAN - LORD CORP. This may be only semantics. I'm disturbed by the use of the word liquid crystals for a polymeric system. Maybe the term mesomorphic phase may be more suitable when talking about polymers than liquid crystals which may be a more proper term for monomeric liquid crystals. F. E. MCFARLANE. As I read the literature on monomeric liquid-crystal materials, I am not sure that they are very well described. Usually one says here are these structures with certain properties and we are going to call them liquid crystals. However, it may be more nearly correct to say we have ordered melts. I feel a little more comfortable with this description, but people will still use the term liquid crystals to describe these materials.

R. NATARAJAN. Have you made any orientation measurements on the moldings? F. E. MCFARLANE. Yes, we have; you can extrude a material such as a fiber with an extremely high degree of orientation as measured by X-ray and wide-line NMR methods.

R. NATARAJAN. Do you have any quantitative results on orientation functions? F. E. MCFARLANE. No, I don't.

R. NATARAJAN. Your viscosity plot as a function of temperature for the para-hydroxybenzoic acid copolymers gave the impression that the nematic phase has a lower viscosity than the isotropic phase. Is this true for all liquid crystal systems?   F. E. MCFARLANE. For all our polymeric systems, in which the viscosity was determined on the flow oriented melts, this is true. I think this is true because in the nematic phase one is orienting domains of molecules rather than individual molecules as in the isotropic phase.

C. F. HAMMER - DU PONT. You showed a figure which included data on heat deflection temperature and molecular weight of molded bars. It appeared that the HDT showed a maximum with increasing molecular weight. Would you care to comment on this?   F. E. MCFARLANE. What you are looking at primarily is the effect of molecular orientation on the heat deflection temperature. The orientation can be a function of both polymer molecular weight and temperature. There may be some temperature-melt viscosity regimes in which, for reasons that could be quite complex, one does not attain the maximum possible orientation.

A. V. POCIUS - 3M CO. Have you used any other polyesters as starting materials besides poly(ethylene terephthalate), e.g., poly(butylene terephthalate)? F. E. MCFARLANE. Yes, we have but it doesn't work quite so well. The reaction scheme requires that the chain scission reaction of the polymer with para-acetoxybenzoic acid be at least competitive with the homo-polymerization of para-acetoxybenzoic acid. For some reason poly(ethylene terephthalate) possesses this characteristic better than any other polymer that we have examined. But if that isn't the case, what you obtain is an inhomogeneous product with homopolymerized poly-p-hydroxybenzoic acid suspended in an isotropic melt.

J. MURRAY - MOBIL CHEMICAL. Your properties went through a maximum with increasing amounts of para-hydroxybenzoic acid. Is there any reason to think that very high relative amounts of para-hydroxybenzoic acid do not form these ordered melts or liquid crystalline states or is this just a function of viscosity effects or melting points?   F. E. MCFARLANE. No, we still have liquid crystals for copolymers containing 80 mole % para-hydroxybenzoic acid. I can change the shape of the curve

by changing the molding temperature. So if one wants to
obtain high physical properties at 80 mole %, just increase
the molding temperature to about 340°C.

R. PORTER - U. MASSACHUSETTS. An historical note -
Glenn Brown who is a U.S. leader in the field of liquid
crystals felt that the subject area should be called the
mesomorphic state and indeed his early review carried this
title, but he subsequently acquiesced to the pressure to
call it liquid crystals. When he founded the only U.S.
institute in the United States devoted to this field he
called it the Liquid Crystal Research Institute.
F. E. MCFARLANE. Thank you Roger. Another point, Professor
Richard Stein (U. Massachusetts) measured the liquid crystal
domain size (if that is what you want to call it) and it
looks as if the domains are roughly 5000 Å, or half a micron,
in diameter. I think Dick will be publishing some of his
data shortly.

K. S. DHAMI - ITT SURPRENANT DIVISION. Regarding the
viscosity, I think it might be very difficult to measure
the viscosity in the perpendicular direction once you apply
some shear. It might orient and again it might flow. When
we are talking about viscosity, of course, I think you mean
the viscosity along the flow direction.   F. E. MCFARLANE.
Yes, the viscosity of the flow oriented melts. There are
many interesting experiments and much chemistry that we have
not explored. We have sent polymer samples to several
people who are interested and who, we hope, will do some of
these experiments. There is a great deal that needs to be
done on the rheology of these polymers. It really holds
the key to many applications. There also remains a great
deal to be done in characterizing these materials.

PROBLEMS IN COMPATIBILITY STUDIES

F. E. Karasz and W. J. MacKnight

Materials Research Laboratory and
Dept. of Polymer Science and Engineering
University of Massachusetts, Amherst MA   01003

I. GENERAL DISCUSSION

For various reasons interest in multi-component polymer systems exhibiting what is variously termed compatibility of miscibility (but usually not solubility) has greatly expanded in the last few years. Though as a result,
the understanding of the properties of such systems has significantly grown, a number of fundamental problems remain.
Several such problems will be discussed, in certain cases
with a degree of unavoidable subjectivity, below. Illustra-
tive examples in nearly all cases will be taken from the
binary system consisting of poly(2,6-dimethyl phenylene
oxide) and polystyrene, and/or derivatives (e.g. copolymers)
of the latter.

Reluctance to adopt the term "solubility" in a discussion of polymer-polymer systems is in itself revealing for
it immediately indicates the first problem we shall consider, the difficulty of establishing a meaningful criterion
(or criteria) for favorable polymer-polymer interactions
leading to compatibility, and the corollary to this, the
question of the molecular conformational state of the polymer which correlate with such a criterion. As is well
known, "compatibility" in polymer-polymer systems has his-
torically been based on strictly pragmatic criteria and it
is only in the most recent past that, in a very few systems,
some progress in understanding the state of the macromolecule on a molecular level has been achieved. Thus the ori-
ginal compatibility criteria--mechanical integrity and (for
amorphous-amorphous systems) optical transparency--have
long been rejected as providing any more than superficial

143

information concerning the conformational state. It is also generally realized that the other criterion commonly employed, the presence of a single glass transition in an
amorphous-amorphous polymer pair, is also not yet meaning-
fully correlated with molecular conformation. There is at
least one report in the literature of a compatible system
which displays a single T_g, calorimetrically determined,
and yet in which electron microscopic studies clearly demonstrate the presence of separate phase domains on a 100-
150A level, (1). A related problem concerns the measurement technique used in assessing the presence or absence of
the single T_g held to denote compatibility. Commonly used
are so-called thermodynamic methods (calorimetry and dila-
tometry) and relaxational techniques (dynamic mechanical
and dielectric relaxation measurements). It is by no means
obvious that in a marginal case all of these techniques
will yield the same result, and indeed there is again at
least one example in which a system appeared to be substantially compatible on the basis of DSC measurements displayed some evidence for phase separation in dynamic mechanical
measurements, (2). A further difficulty arises when this
criterion is applied to binary systems in which the T_g's of
the constituents are relatively close.

An additional problem concerning T_g determinations is
related to the fact that such measurements necessarily reflect the state of the system at a temperature substantially
removed than T_g itself. An increasing number of systems is
being found that display a lower consolute point and in
which, therefore, depending on the thermal history of the
sample, phase separation may or may not have occurred. In
such instances it will be possible to observe single or
double T_g's in binary samples of identical composition. A
related observation concerns certain binary systems whose
preparative history may have included removal of a solvent
by precipitation from, or evaporation of, the low molecular
component(s). In these cases the phase state of the polymer
mixture presumably reflects the detailed path from dilute
solution to solid sample.

The general difficulty of attaining an equilibrium
state stemming, clearly, from the long relaxation times associated with polymer melts, is also reflected in the vir-
tual absence of thermodynamic data for polymer-polymer mixtures in terms of solubility theory. For example very few
pertinent heats of mixing have been reported and there is

no complete thermodynamic analysis as yet available for any
system in which both components are of high molecular weight,
(3). A number of interaction parameter studies have appeared, the data mostly obtained from ternary polymer-polymer-
solvent studies, (4), but a general correlation of such
parameters (using well established polymer-solvent precepts)
with compatibility is not available. This then brings us
to a final problem--the prediction of mutual solubility in
polymers. It is well known that the latter is a relatively
rare phenomenon, and a simple explanation of this fact is
associated with the difficulty of obtaining the requisite
negative free energy of mixing for any polymer-polymer pair.
(A negative ΔG_m is a necessary but not sufficient condition
for the formation of a single homogeneous phase). It is
apparent from experimental results on several systems that
the free energy balance is a delicate one and that in many
if not all cases a small change in the chemical structure
of either or both components of a compatible system is sufficient to cause phase separation. It may be presumed that
this criticality is accentuated by an increase in the molecular weight of the constituents (though it may in this
limiting case be masked by non-equilibrium effects). Thus
any a priori attempt to predict mutual solubility on the
basis of the properties of the proposed components is in-
herently a formidable one. None of the methods, empirical
and semi-empirical, which have in the main been successful
for predicting polymer solubility in potential low molecular
weight solvents, have been found to be applicable to the
polymer-polymer case. Some progress has been achieved using equation of state concepts, (5), but it appears here
again that physical parameters such as bulk thermal expan-
sivity which may be presumed on theoretical grounds to have
some predictive value in this regard will require "matching"
to a degree which is beyond their experimental determinabi-
lity. It would seem that polymer-polymer solubility is
necessarily as yet explored on a somewhat empirical basis.

II. A MODEL SYSTEM:
POLY(2,6-DIMETHYL-1,4-PHENYLENE OXIDE)-POLYSTYRENE

Introduction. A system with which most of the problems
and effects discussed above may be rather effectively illustrated is poly(2,6-dimethyl-1,4-phenylene oxide)-polysty-
rene (PPO-PS), and systems derived from this. In the following section we shall briefly review the principal results

obtained by the authors and co-workers relating to this
versatile system.

1. Order. PPO-PS is unique amongst all known compatible polymer pairs in that it is possible to obtain, by
appropriate selection of the tacticity of the PS component,
and of the thermal and solvent exposure history of the in-
timately blended components, blends which are either totally amorphous (using atactic PS), in which one or other of
the two components are partially crystalline (using PPO and
isotactic PS) or, finally, in which both components (PPO,
i-PS) exhibit substantial degrees of crystallinity. Conditions necessary to obtain each of the conditions delineated
above have been explored in detail, (2,6). A principal
area of interest in the systems containing either or both
components in discrete (and, of course, separate) crystalline regions concerns the melting points of the respective
crystallites. In systems considered here substantial melting point depressions were observed as the fraction of "diluent" increased. Such depressions are predicted to occur
in polymer-polymer systems in recent theoretical treatments
of this problem, (7), (though it was originally believed
that a high molecular-weight diluent should be incapable of
causing such a decrease). Nevertheless it is clear that a
melting point depression in a polymer-polymer system can be
a consequence of either or both thermodynamic and kinetic
effects. The former would result essentially from free
energy decrease upon mixing, the latter from the potential
failure of the crystallites to attain substantial size. In
the PPO-i-PS system wide- and small-angle x-ray studies
were conducted to further investigate this point. The analysis of these results, using as a basis a paracrystalline
model, for blends in which only the i-PS had been allowed
to crystalline found that the lamellae thickness of the
i-PS component did indeed decrease with the increasing PPO
content, thus explaining, in part, the observed m.p. depression, (8). The results of the complementary study, in
which m.p.'s are determined as a function of annealing conditions (and in which the effect of crystal size can, therefore, in effect, be removed from consideration by extrapolation) are not yet complete in this system.

2. Chemical Modifications and Compatibility. The PPO-
PS system (using either atactic or amorphous isotactic PS)
is compatible, by all known criteria, over the entire com-

position range. However, and in parallel with empirical observations in other compatible binary mixtures, relatively
small changes in chemical structure of either polymer component are sufficient to render the resulting system entirely incompatible. For example, both the 2-methyl-6-phenyl
or the 2,6-diphenyl phenylene oxide polymers are incompatible in mixtures with PS, (9). Similarly neither p-chloro
nor the o-chlorostyrene homopolymers are compatible with
PPO. However, certain copolymers within a strictly defined
range of composition are compatible; thus p-chlorostyrene
(pClS) and styrene copolymers containing up to 67 mole %
pClS are compatible. The compatibility-incompatibility
transition in the latter system has received extensive study
with respect to solid state transitional and mechanical properties. The incompatibility transition in the copolymer
is an extremely sharp one, with a half-width of the order
of 1 mole %. The same criticality is exhibited with respect
to copolymer-PPO blends. Thus, as far as is experimentally
ascertainable a given compatible copolymer is compatible in
all proportions with PPO; for any incompatible copolymer
the converse is true, (10).

Further consideration of the sharpness of the compat-
ible-compatible transition along the copolymer composition
axis, however, needs to be made taking into account the possibility of non-equilibrium effects. It is recalled that
the criterion of a single T_g necessarily reflects a thermodynamic pseudo-equilibrium at some temperature other than
the measured T_g itself; in samples prepared by melt blending
presumably at temperatures close to or at the processing
temperatures. This important point is illustrated by recent
observations concerning the compatibility of p-ClS and o-ClS
copolymers with PPO. A compatibility "window" exists in
this system, but the width of the "window" is highly dependent on the thermal history of the sample. It has been deduced that a lower critical consolute point exists for mixtures
of the copolymer and PPO. For the "most compatible" copolymer i.e. with compositions near the center of the "window",
the critical temperature is substantially higher than any
normally employed processing temperature, hence all blends
exhibit compatibility. However, for copolymer compositions
nearer to the edges of the "window", the LCST curve inter-
sects the processing temperature regions. Thus samples of
a given composition with respect to copolymer and PPO
quenched from a higher temperature will exhibit incompati-

bility, whereas samples processed at an appropriately
lower temperature appear to be compatible. The fullest exploration of the edges of the transition is compromised, of
course, by rapidly increasing equilibrium times as the temperatures of interest approach the respective T_g's, (11).
These studies provide further evidence of the subtlety of
the compatibility--incompatibility phenomenon with respect
to chemical composition for high molecular weight components and the evident difficulty in accounting for such effects in terms of any bulk properties of individual constituents. Although in the pClS-PS copolymer system we have
been able to correlate compatibility with respect to mixtures with PPO with a densification of the respective compatible copolymers (of the order of 1% at most, compared
to a simple additivity prediction), some preliminary analogous measurements in the pClS/oClS copolymer compatibility
"window" have failed to show such an effect.

3. **The Glass and Other Solid State Transitions.** The
PPO-PS system and compatible derivatives thereof all exhibit what has become the standard criterion of miscibility
in polymer-polymer blends: a single glass transition temperature coupled with the absence of any manifestation of
the T_g's of the component polymers. Additionally, the
single T_g varies in a momotonic manner with the composition of the blend, and may be functionally defined by any
of the semi-empirical relations that have been derived to
account for the composition dependence of T_g for a polymer-
diluent mixture. Somewhat more subtle aspects are revealed
when a comparison is made of the T_g phenomenon observed by
different experimental techniques. Thus in an early investigation of this system it was found that in PPO-PS
blends prepared under a given set of conditions (not necessarily representing "asymptotic" equilibrium behavior) a
single T_g was observed in DSC measurements whereas some
evidence of less than complete miscibility in the same
samples could be inferred from dynamic mechanical considerations, (2). The resulting question, could there be a
difference in the sensitivity to compositional inhomoge-
neity in the two respective techniques, has not been defi-
nitively answered and this, together with other aspects of
glass transition behavior, remain important unknown factors
in the comprehensive understanding of the complete phenomenon.

The effect of compatibility on solid state relaxations

other than these associated with the glass transition in
PPO-PS is somewhat obscured by the coincidental overlap of
the major sub-T_g relaxations in the two components. The
respective β-peaks are both broad and, in the dynamical
mechanical domain, relatively weak. There is some evidence
that blending causes a suppression of the β peak and this
has been attributed to a specific interaction, (12), but
it seems that this particular phenomenon may be more fruit-
fully investigated in a compatible system in which the
sub-T_g relaxations in the individual components occur at
substantially different temperatures.

Finally we discuss in this section the special problem of assessing compatibility in amorphous binary mixtures in which the T_g's of the individual constituents
happen to lie comparatively close. A possible solution
to this problem lies in observations, now confirmed in
both PPO-PS and PPO-pClS/PS copolymers, regarding large
deformation properties of such blends. It has been found
that these properties, particularly the tensile strength,
offer a sensitive additional criterion. Thus for compatible systems a pronounced maximum is observed in measurements of tensile strength as a function of blend composition. In incompatible systems, conversely, a minimum is
found, (10,13). Additional tests of the general applicability of this suggested criterion particularly with respect to brittle-brittle polymer systems is under way.

4. Level of Miscibility. A question of prime interest in all compatible systems naturally concerns the level
of dimensional homogeneity in such a system. Although it
is possible to assert from electron microscopic observations, for example, that discrete phases with an average
characteristic dimension of greater than, say, 50-100A are
not likely to be present in PPO-PS blends (or, most probably, in any system displaying a single T_g) experimental
limitations (electron density contrast, for example) do
not permit any more definitive statement in this regard.
Neither, as has been mentioned above, is it known, how the
single T_g phenomenon may be correlated with dimensional
homogeneity. Whether or not a "true" solution in terms
of a macromolecular--low molecular weight solvent concepts
exists in polymer-polymer systems has not been readily
ascertainable. Recent results from small angle neutron
scattering measurements have provided evidence, for those
systems examined, which strongly support the concept of a

molecular, segmental, level solution in which the R.M.S.
end-to-end distances of the chains of the constituents are
mutually expanded with respect to their respective unperturbed values, a situation, therefore, to be anticipated
for any polymer in a thermodynamically favorable solvent,
(14). We have reached a similar conclusion in the compatible PPO–pClS/S copolymer system from dielectric relaxation measurements, which provide, again, strong evidence,
in this particular system, for an extended pClS/S chain in
the PPO "solvent", (15).

Acknowledgement. This work was supported in part by
AFOSR Grant 76–2983.

REFERENCES

1. M. Matsuo, S. Sagae and H. Asai, Polymer 10, 79 (1969).
2. J. Stoelting, F.E. Karasz, W.J. MacKnight, Polym. Eng.
   Sci., 10, 133 (1970); ibid., in Multicomponent Polymer
   Systems, A.C.S. Washington, D.C., 1971, p. 29.
3. M.P. Zverev et al. Vysokomol. soyed., A16, 1813 (1974);
   S. Ichihara, A. Komatsu, T. Hata, Polym. J., 2, 640
   (1971).
4. A.R. Shultz and C.R. McCullough, J. Polym. Sci. A2, 10,
   307 (1972).
5. P.J. Flory, B.E. Eichinger, and R.A. Orwoll, Macromol.,
   1, 287 (1968); J.P. McMaster, Macromol., 6, 760 (1973);
   I.C. Sanchez and R.H. Lacombe, J. Phys. Chem, XXX 80,
   (1976).
6. R. Neira, F.E. Karasz and W.J. MacKnight,
   in press.
7. T. Nishi and T.T. Wang, Macromolecules, 8, 909 (1975).
8. W. Wenig, F.E. Karasz, and W.J. MacKnight, J. Appl.
   Phys., 46, 4194 (1975); R. Hammel, W.J. MacKnight, F.E.
   Karasz, J. Appl Phys., 46, 4199 (1975).
9. A.R. Shultz and B.M. Gendron, J. Macromol. Sci. Chem.
   Ed., A8, 175 (1974).
10. J. Fried, W.J. MacKnight and F.E. Karasz, in press.
11. P. Alexandrovitch, F.E. Karasz and W.J. MacKnight, J.
    Appl. Phys., 47, 4251 (1976), and in press.
12. E. Wellinghoff and E. Baer, Preprints. Org. Coat. Plast.
    Chem., 36, 140 (1976).
13. L. Kleiner, F.E. Karasz and W.J. MacKnight, in press.
14. W.A. Kruse, R.G. Kirste, J. Haas, B.J. Schmitt, D.J.
    Stein, Makromol. Chem., 177, 1145 (1976).
15. J. Tkacik, W.J. MacKnight and F.E. Karasz, in press.

DISCUSSION

N. DODDI - ETHICON, INC. The subject of polymer compatibility you discussed here, can you extend this to processing of polymer mixtures? Are they compatible from the processing point of view? Can we use this information presented by you to predict the processibility of polymer mixtures? F. KARASZ - U. MASSACHUSETTS. That's a many faceted question. First of all to answer the simple part. Yes, if a material is compatible, and if one knows the phase boundaries, the kind of information I gave can give you an indication of what temperature one should process. For example, in the p-chlorostyrene/o-chlorostyrene/PPO system that I discussed, we have a compatibility window. Clearly if one were interested in developing a material which was compatible ultimately at room temperature one would process at a temperature A. If, on the other hand, for reasons of one's own, there was interest in producing a system consisting of a matrix containing discrete phases one would process in the range B and cool quickly. This kind of diagram thus does indeed give one potentially very important information about processing conditions for the desired end product. As I said, it might well be that for other systems the L.C.S.T. is either nonexistent or is at such a high temperature that it doesn't make any difference. One would have to examine each situation individually. See Phase Diagram Figure on following page.

ELI PEARCE - POLYTECHNIC INSTITUTE OF NEW YORK. I'm wondering whether you've also given any consideration to the variation of copolymer structure with conversion. I believe when you prepared your copolymers they were run to rather high conversions. Therefore, they are really a polymer "blend" themselves, and how does this affect the results? Perhaps if low conversion copolymers were used, such difficulties would be minimized. F. KARASZ. Yes, we were very aware of this problem and the potential dangers. It just happens that in the particular copolymer that we used, reactivity ratios are such that product composition changes only 1% with conversions from 20 to 60%. We showed this experimentally and theoretically. In other cases it might be more crucial.

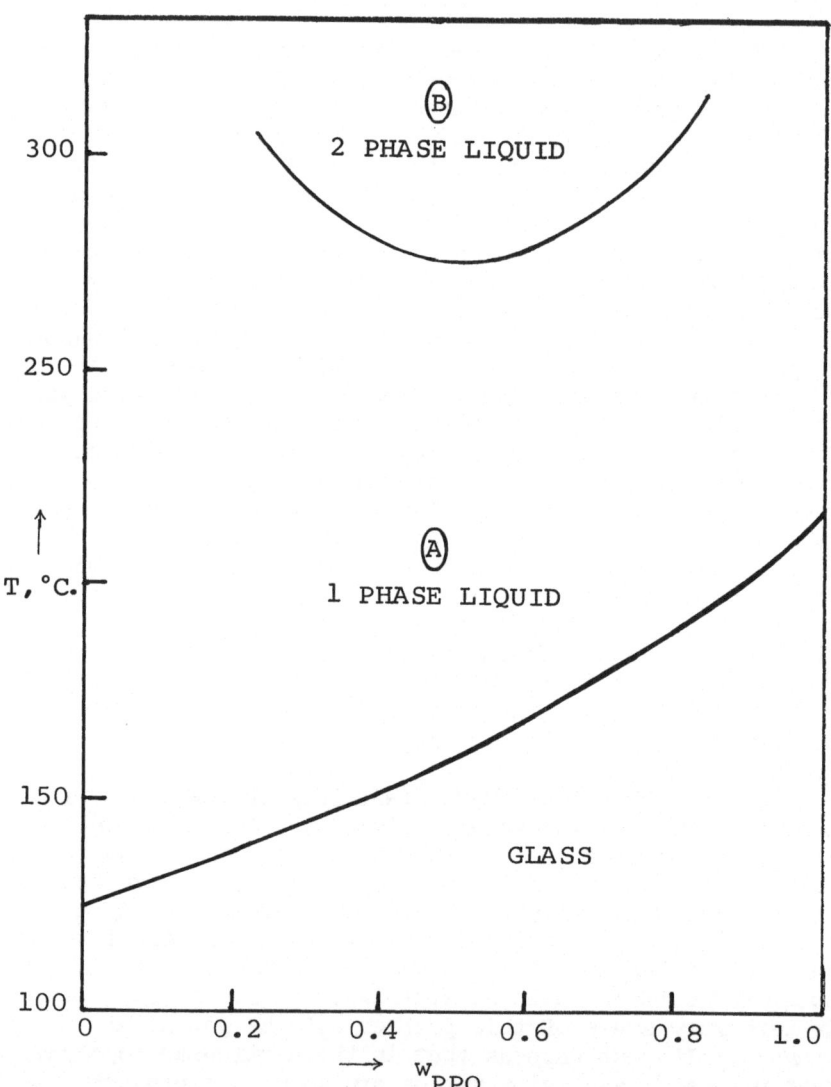

Phase diagram for a blend of a p-chlorostyrene/o-chlorostyrene copolymer (p-chlorostyrene content 47 mole %; $\overline{M}_V$ = 2.2 x 10^5) with poly(2,6-dimethyl-phenylene oxide) as a function of weight fraction (w_{PPO}) of the latter polymer.

J. LANDO - CASE WESTERN RESERVE U. Have you given any consideration to the fact in your tensile experiments that as you vary the compositions you may get a different mechanism for tensile failure?   F. KARASZ. Yes. As the amount of PPO increases there is a transition from brittle to ductile failure.

J. LANDO. One last point. Have you ever thought of using corresponding fluorinated polymers because then you could perhaps separate the effect on compatibility of size and polarity?   F. KARASZ. Yes, we are starting to look at such systems.

T. W. HUSEBY - BELL LABS. My comment refers to the previous question on processibility, and I wanted to reinforce what Frank had said about predicting processibility from some of these studies. In our own work we have found compositions of PPO and polystyrene in which the zero shear viscosity scales with concentration of polystyrene in a very predictable manner and in which the shape of the viscosity-shear rate curve also scales in a very predictable manner.

RICHARD CHEN - BAUSCH AND LOMB. In your presentation, it is interesting to see the correlation between T_g's and the compatibility of two blended polymers. I wonder, do you do any work using electron microscopic sectioning techniques to correlate the domain size of blended polymers and the T_g's?   F. KARASZ. Yes, with the exception of the last system I discussed: the o-chloro and p-chlorostyrene where we have not yet examined the edge of compatibility. With the other systems, i.e., the PPO-polystyrene and the PPO-p-chlorostyrene/styrene copolymers we have looked at the systems electron microscopically. We do not observe anything that is inconsistent with T_g measurements. In other words we do not observe an effect such as I mentioned Matsuo had found where he saw discrete phases and yet found a single T_g. In other words, in our case we go from a discrete phase to a homogeneous phase situation at exactly the same point as our T_g measurements indicate. All that means, of course, is that we are possibly too insensitive in our visual observations to the subtleties of the transitional region.

RICHARD CHEN. Some people have been working on the
so-called interpenetrating polymer network (IPN) systems.
There have been some arguments about the T_g values too.
Do you think they should obtain the separate T_g's as you
said, even if the IPN samples are transparent? How small
should the domain size be in order to obtain a single T_g?
F. KARASZ. It depends on two things. (1) How discrete are
the phases? The components are presumably not compatible
in the sense that I am talking about. They are forced into
proximity but not compatibility. (2) It also depends on
the answer to the fundamental point you raised. How small
a region do you have to have before it shows a single T_g?
There is evidence that a region as small as 50 or 60 Å can
show a single T_g but if you get down to 10 Å (even if that
has a real meaning) I don't know whether that would show a
single peak.

I. SKEIST - SKEIST LABORATORIES. With regard to
predicting compatibility from the properties of the constituents, how useful are solubility parameters, either the
single value suggested by Hildebrand or the triple values
of Hansen and the Tennessee Eastman people (Crowley and
coworkers)?   F. KARASZ. Well, it is perhaps good for
predicting noncompatibility, but, in my opinion, not very
productive for predicting compatibility simply because the
subtleties of the problem are a couple of orders of magnitude
greater than those reflected in the solubility parameter.
For example, if we talk about, as a very obvious example, the
systems I discussed, any method of predicting solubility
parameters that I know for the p-chloro and o-chlorostyrene
copolymers will give essentially the same value. There is
no way, therefore, in which a solubility parameter argument
will predict the observed compatibility window in such a
system.

R. PORTER - U. MASSACHUSETTS. T_g is, of course, just
one position on the viscoelastic spectrum and in a study
that you know about, Frank, we have shown that the full
viscoelastic spectrum is additive for the PPO-polystyrene
system, which is germane to the comment by Dr. Huseby that
viscosities are additive, so that we can literally predict
the rheological properties as long as we are working in an
area where the two materials are fully compatible.

FRED BAILEY - UNION CARBIDE.   One system on which I
have worked is the p-chlorostyrene/styrene isotactic
copolymer system.   Here the isotactic conformation can be
made but only the styrene portion will crystallize; the
chlorostyrene segments are amorphous.   Thus, the degree of
crystallinity can be closely controlled.   The chlorostyrene
dipoles can then be measured in a system which is crystalline:
polystyrene-PPO and varying degrees of polystyrene-poly-p-
chlorostyrene crystallinity.

COMPATIBILITY AND PHASE SEPARATION IN POLYMER MIXTURES

T. K. Kwei

Bell Laboratories

Murray Hill, NJ 07974

The thermodynamic criterion for a stable, single-phase, binary mixture is that the second derivative of the free energy with respect to concentration must be positive. Although the criterion is universal for molecules of all sizes, many standard techniques of measuring properties of mixtures of simple liquids become insensitive for high molecular weight, solid polymers. Instead, it is often necessary to use somewhat arbitrary working definitions of compatibility based on measurable characteristics of the polymer systems. Several of the commonly used methods are light transmission, microscopy, scattering of light or X-ray, thermal expansion, heat capacity and dynamic techniques such as dielectric, methanical, and nuclear magnetic relaxations. Each of these measuring techniques has its advantages and limitations in resolution and interpretation. Therefore it is desirable to apply many different approaches to examine the diverse aspects of the homogeneity of mixing of segments.

An additional complexity is the existence of two critical temperatures. When the temperature of an initially homogeneous mixture is lowered below the phase boundary, phase separation begins to occur. The temperature of phase separation is dependent on composition and the critical point associated with the cloud point curve is called upper critical solution temperature. (UCST) For some polymer-solvent and polymer-polymer mixtures phase separation is also observed when the temperature of the mixture is raised.[1-4] The corresponding critical point is called lower critical solution temperature (LCST), although it is actually at a higher temperature than UCST.

The experimental identification of the phase boundary is further impeded by the high viscosity of the polymeric medium which governs the rate of phase separation. If the glass transition temperature, Tg, of the mixture exceeds the UCST, kinetic considerations are likely to prevail and the extent of phase separation may be influenced strongly by thermal history.

In this paper a brief review is given for our studies of three types of blends, namely, two amorphous polymers, a crystalline-amorphous combination, and a thermoplastic elastomer with an amorphous polymer.

1. In the first category, polystyrene (PS) and poly(vinyl methyl ether) (PVME) have been reported to be compatible when films are cast from suitable solvents.[5] But phase separation can be induced in a compatible film simply by heating the mixture to about 150°C.[6] Subsequent slow cooling of the phase-separated mixture results in a compatible film. The thermally induced phase separation behavior of PS-PVME mixture indicates the existence of LCST. The presence of UCST was not reported, although, according to Patterson's theory,[3] if LCST exists, UCST should also exist.

Our investigations[7-9] of this system center on four aspects. First, to establish that the compatible mixtures indeed represent the thermodynamically stable states at room temperature; second, to find the upper phase boundary of the mixtures; third, to explore the use of nuclear magnetic resonance to study molecular motion which allows us to draw conclusions about the scale of homogeneity of mixing; and fourth, to study the morphology and kinetics of phase separation, in particular, spinodal decomposition associated with LCST.

(a) The thermodynamic study will be presented first because it has some bearing on the detection of the upper phase boundary. Since there is no practical experimental method to measure the thermodynamic interaction between two solid polymers directly, we make use of the polymer solution theory applied to ternary systems consisting of one solvent and two polymers. The activity a_1 of the solvent in a ternary solution is[10]

$$\ell n \; a_1 = \mathfrak{m}\phi_1 + (1-\phi_1) + (\chi_{12}\phi_2 + _{13}\phi_3)(1-\phi_1) - \chi_{23}'\phi_2\phi_3 \quad (1)$$

where the subscript 1 refers to the solvent and subscripts 2 and 3 refer to the two polymers. The volume fraction of each component is represented by ϕ and the binary inter- action parameter is given by χ. The symbol χ_{23}' is the interaction per segment of polymer 2, i.e., χ_{23}/m_2 where m_2 is the number of segments in the macromolecule.

Since the solvent acts only as a probe to permit the use of eq.(1) we choose to conduct our study with only a minor amount of solvent in the ternary system so that the information obtained from our experiments may reflect closely the property of a binary polymer mixture. The technique of our choice is the vapor sorption measurement in which the polymer is equilibrated with an organic vapor at a predetermined vapor activity and the weight gain is measured by an electrobalance. The parameters χ_{12} and χ_{13} can be determined from separate experiments, (benzene, PS, PVME as components 1, 2, and 3 respectively),

$$\ell na_1 = \ell n \phi_1 + \phi_2 + \chi_{12}\phi_2^2 \qquad (2)$$

With a knowledge of χ_{12} and χ_{13}, χ_{23}' can be readily cal- culated from the vapor sorption data of the blends. The computed values of χ_{23}' are shown in Fig. 1.

We note in our measurements that χ_{23}' does not change significantly with the concentration of the solvent in the ternary system. The values reported here are there- fore applicable to blends without solvent. Secondly, the negative values of χ_{23}' is a sufficient condition, though not a necessary one, which satisfies the criterion of thermodynamic stability for binary polymer mixtures. The third point of interest is the dependence of the inter- action parameter on concentration. Extrapolation of our experimental data suggests that χ_{23}' would become positive at 70% PVME and compatibility might be marginal. Indeed, a film of this composition was initially transparent but turned opaque after standing at room temperature for several months. A mixture containing 75% PVME became opaque after three weeks. The opaque film after being heated at 47°C for 40 hours, turned transparent again. Upon further elevation of temperature to 150°C, the film once again became opaque and NMR data indicated conclusively

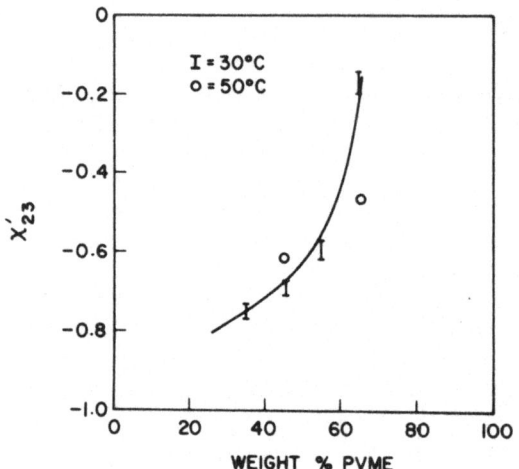

Figure 1

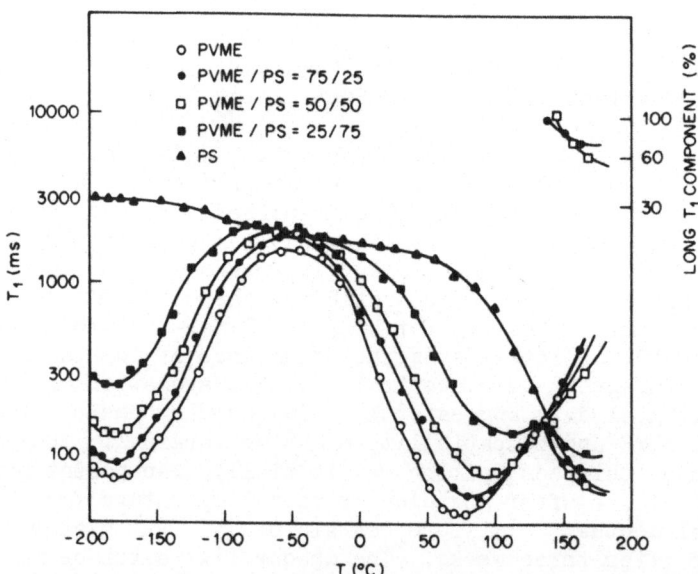

Figure 2

that phase separation had taken place. Thus we have located the upper phase boundary for at least one composition.

(b) The application of NMR to the study of polymer mixtures offers the attractive possibility in that phase separation may be manifested by the appearance of multiple relaxation times. In Fig. 2 the temperature dependence of spin-lattice relaxation time T_1 for PVME, PS, and their mixtures are shown. Only one T_1 was observed for every sample except for PVME/PS = 75/25 and 50/50 above 140°C.

The high temperature T_1 minimum at 75°C for PVME is associated with glass transition. The increase in the temperature of T_1 minimum with PS content in the mixture parallels the trend observed in DSC studies.

In the low temperature region, the T_1 minimum at about -175°C is associated with methyl group rotation in PVME, and its position remains unchanged in the mixtures. The value of T_1 minimum correlates linearly with the number of methyl protons in the mixtures. (This correlation can be used to calculate the composition of the phase in a phase-separated mixture to be discussed later.) The same correlation, found in the study of n-alkanes, was interpreted as an indication of effective energy transfer from methylene to methyl groups through spin diffusion of the proton. Similarly, we conclude that the two polymer chains are mixed quite thoroughly, at least within the distance of effective spin diffusion which is about 30 to 50Å. We now have an estimate of the scale of homogeneity.

The spin-spin relaxation times are shown in Fig. 3. Only one T_2 was observed at low temperatures but two were found for the 50/50 mixtures above 25°C. The interpretation of multiple T_2 is not completely unambiguous but it is known that T_2 measurements is more sensitive for the detection of multicomponent systems than T_1 because it is free of spin-diffusion effects. Taken at face value, the T_2 results suggest that heterogeneity may exist but on a scale small compared to the distance of effective spin diffusion. But an improvement in interpretation is clearly desirable.

At temperature beyond 140°C, the emergence of two spin-lattice relaxation times is a gratifying demonstration of

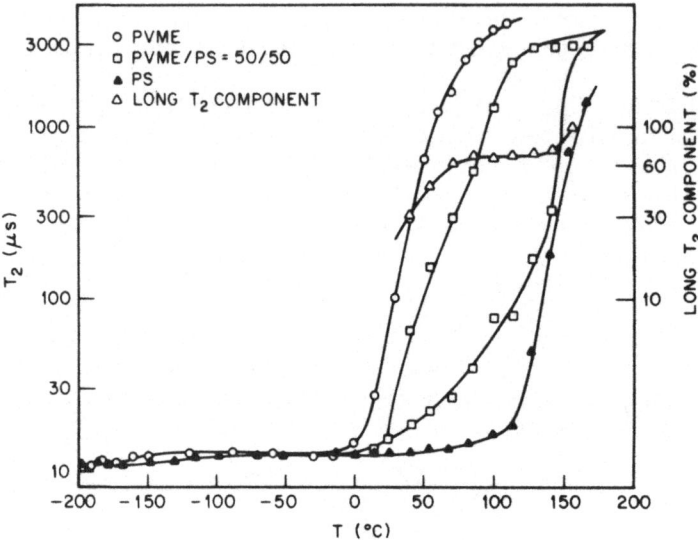

Figure 3

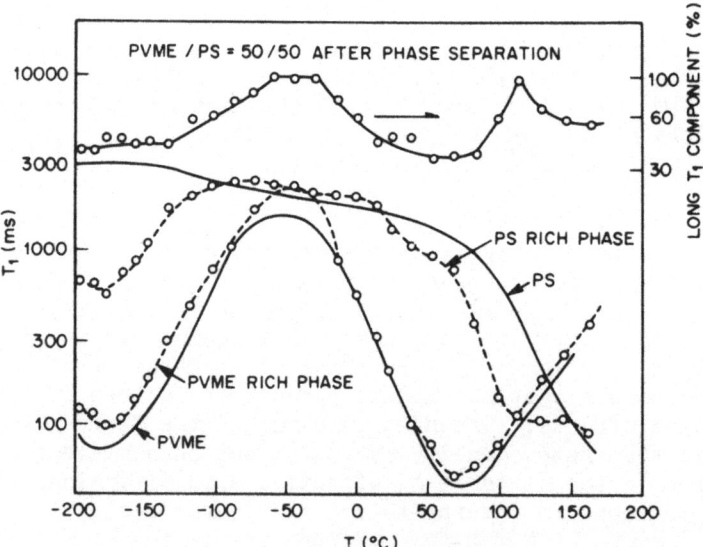

Figure 4

the usefulness of the NMR method in studying phase separation. After NMR measurements were completed above 160+C
the phase-separated samples were quenched in ice-water and
measurements were carried out again. Typical results for
a 50/50 sample before and after phase separation are shown
in Fig. 4. Two relaxation times are evident for the phase
separated film at all temperatures. From the correlation
between the value of low temperature T_1 minimum and the
number of methyl protons, the PVME concentration in the
two phases are calculated to be about 10 and 70% respectively.

(c) Two mechanisms of phase separation can be operative,
namely, nucleation and growth in the metastable region and
spinodal decomposition in the unstable region where the
second derivative of free energy with respect to concentration is negative. The three important features of Cahn's
theory[11] of spinodal decomposition are the following:

(i) In the early stages of spinodal decomposition
the composition of phases change continuously with time
while phase volumes remain nearly constant.

(ii) The diffusion coefficient must be negative
(uphill diffusion).

(iii) For isotropic materials, the spinodal structure
is almost uniform but randomly interconnected.

Experimental observation of the morphological features
of phase separated films confirms the existence of two
mechanisms. In addition, NMR measurements provides information about phase compositions and phase volumes as spinodal decomposition proceeds (Table I). The column for
percent long T_1 components represents the volume of PS-
rich phase. The values indeed remain nearly constant,
as predicted by Cahn, at least for the first four minutes.
The last column illustrates the continuous decrease in PVME
content with time.

Together with thermodynamic and morphological information, the NMR data allows the calculation of all the basic
parameters governing the kinetics of spinodal decomposition.

2. In the study of blends of a crystalline polymer
with an amorphous one, the use of the vapor sorption method

to obtain interaction parameter no longer offers the same
advantage when applied to amorphous - amorphous combination.
The uncertainty in the degree of crystallinity of the
crystallizable component in the mixture often invalidates
the calculation. Instead, an alternate approach based on
the depression of the melting point of the crystallizable
polymer appears to be successful.[12] When the change in
the chemical potential of the crystallizable segment in
the mixture relative to its pure liquid state is equated
to the difference in the chemical potential between a
crystalline segments and the same segment in the pure
liquid state, the following equation results:

$$\frac{\Delta T_m}{T_m^{\,o}} = - \frac{R \, \bar{V}_{2u} \, T_m}{\Delta H_{2u} V_{1u}} \quad \chi_{12}' \, \phi_1^{\,2} \qquad (3)$$

In the above equation, $T_m^{\,o}$ is the melting temperature of
the undiluted crystalline polymer (component 2), ΔT_m is the

Table I

T_1 at 50° for PS-PVME = 50:50 Compatible Polymer Mixtures
Decomposed at 130° for Various Lengths of Time

Decompo-sition time	T_1		Long T_1 compo-nent, %	PVME % in PS-rich phase
	Long T_1, msec	Short T_1, msec		
Original		195		
20 sec	390 ± 30	115 ± 15	40 ± 8	29.5 ± 1.5
40 sec	375 ± 20	130 ± 10	35 ± 6	30 ± 1
1 min	420 ± 20	120 ± 10	37 ± 6	28 ± 1
2 min	410 ± 10	115 ± 5	39 ± 4	27.5 ± 0.5
4 min	530 ± 10	112 ± 5	38 ± 4	23.5 ± 0.5
8 min	610 ± 15	120 ± 5	31 ± 4	21.5 ± 0.5
32 min	680 ± 15	115 ± 5	33 ± 4	19.5 ± 0.5

lowering of the melting point, $\bar{V}_{1u}$ and $\bar{V}_{2u}$ are the molar volumes per segment, and ΔH_{2u} is the heat of fusion per crystalline segment. It is seen from eq.(3) that the depression of melting point can be realized only if χ'_{12} is negative. Furthermore, ΔT_m is directly proportional to magnitude of the interaction parameter.

Expressing the interaction parameter as

$$\chi'_{12} = A_{12} + B_{12}\,\bar{V}_{1u}/RT \tag{4}$$

one obtains, upon substitution, eq.(5),

$$\frac{1}{\phi_1^2}\frac{\Delta T_m}{T_m^{\,o}} = \frac{R\,\bar{V}_{2u}}{\Delta H_{2u}V_{1u}}\;(A_{12}T_m + B_{12}\bar{V}_{1u}/R) \tag{5}$$

A plot of $\Delta T_m/\phi_1^2 T_m^{\,o}$ v.s. T_m should result in a straight line from which A_{12} and B_{12} can be calculated if all the physical constants are known. Such a plot, however, magnifies the error of $\Delta T_m/\phi_1^2 T_m^{\,o}$ when ϕ_1 is small and the uncertainty in ΔT_m is large. Rather, it is more convenient to use a simplified form of Eq.(5) when A_{12} is small compared to $B_{12}\bar{V}_{1u}/R$,

$$\frac{1}{\phi_1}\frac{\Delta T_m}{T_m^{\,o}} = -\frac{\bar{V}_{1u}\,B_{12}}{\Delta H_{2u}}\;\phi_1 \tag{6}$$

and to make use of the plot of $\dfrac{1}{\phi_1}\dfrac{\Delta T_m}{T_m^{\,o}}$ v.s. ϕ_1.

The melting point data of poly(vinylidene fluoride) in the presence of poly(methyl methacrylate) or poly(ethyl methacrylate) obey the prediction of eq.(b), as shown in Fig. 5. The B_{12} values are -2.98 cal/cm^3 PMMA[12] and -2.86 cal/cm3 PEMA.[13] The addition of a methylene group to the side chain of the methyl methacrylate polymer does not seem to have significant influence on its compatibility with PVF$_2$ at elevated temperatures.

The possibility exists that the depression in melting point may also arise from morphological changes such as imperfections in the crystals and reduction in lamella

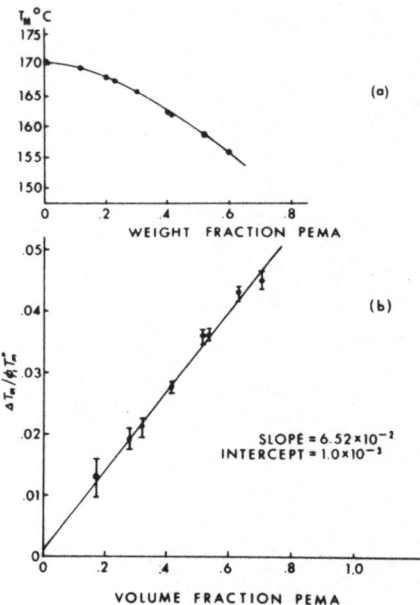

Figure 5

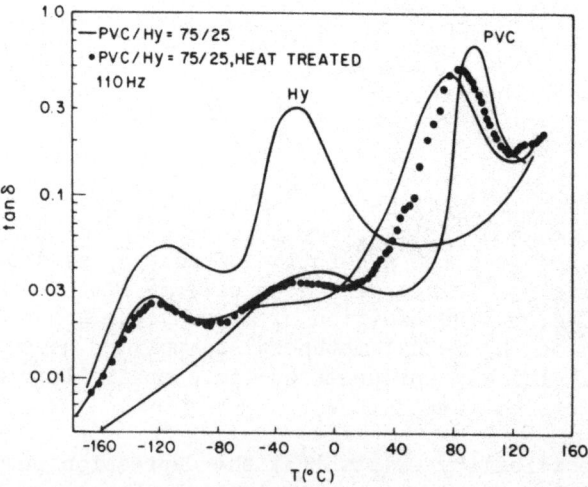

Figure 6

thickness. However, some clues can be found in the fusion
characteristics of isothermally crystallized samples of
PVF_2-PMMA. A linear relationship was found between T_m
and T_c and the stability parameter, which was in fact a
morphological parameter, was independent of composition.
This suggests that morphological changes are not major
contributions to the lowering in melting point.

Quenched films of PVF_2 and PMMA exhibit a single T_g
for each composition and the value of T_g is simply the
volume fraction average of the T of the component polymers.
Quenched specimens containing 60% or more PEMA also exhibit
single glass transitions. When the PEMA content is below
40% crystallization of PVF_2 prevails even at high cooling
rates. In addition, two conjugate amorphous phases, one
consisting of nearly 100% PVF_2 and the other about 45%,
are present.

The above study was extended to ternary mixtures of
PVF_2, PMMA, and PEMA. Although both methacrylate polymers
are compatible with PVF_2 and the binary interaction parameters are similar in magnitude, they are incompatible
with each other. Therefore we explored the possibility of
rendering the two methacrylate polymers compatible by using
PVF2 as a common solvent.[14] The melting point depression
of PVF_2 in the ternary system can be calculated from Scott's
equation. Mixtures containing 40 to 70% of PVF_2 are amorphous when quenched from melt and each mixture consists of
a single phase whose T_g is equal to the volume fraction
average of the T_g of component polymers.

3. The third system[15] of our investigation consists
of poly(vinyl chloride) and a copolymer formed by randomly
joining poly(tetramethylene ether) glycol terephthalate,
the soft segments, and tetramethylene terephthalate, the
crystallizable hard segments. Only one spin-lattice relaxation time was observed for each mixture, indicative
of extensive mixing of different types of segments. At
the same time, dynamic mechanical properties (Fig. 6)
appear to reveal pertinent clues about the degree of
mixing. In the temperature region from -60 to 0°C where
the glass transition of the copolyester and the local motion
of PVC segments occur, the loss tangent curve of the blend
containing 75% PVC is almost flat. At higher temperatures
a single loss maximum associated with the glass transition
of the mixture is prominent. The appearance of a single

major glass transition and the change in the secondary
relaxation of PVC are strong evidences that the local
environment of PVC segments has been altered by the
presence of polyester segments.

The microstructure of the melt-mixed blend, however,
was dependent on the thermal history of the specimen. By
an appropriate heat treatment at 130°C, it underwent phase
separation[15] with an attendant seven fold increase in
room temperature impact strength.[16] The mechanism of energy
dissipation involves both crazing and shear flow processes.
These observations suggest the existence of UCST. The
phase domains in the annealed samples are likely to be
extremely small.

REFERENCES

1. P. I. Freeman and J. S. Rowlinson, Polymer, I, 20(1960).
2. D. Patterson, Macromolecules 2, 672 (1969).
3. G. Delmas and D. Patterson, Official Digest, 1, (1962).
4. S. Konno, S. Saeki, N. Kuwahara, M. Nakata, and
    M. Kaneko, Macromolecules 8, 799 (1975).
5. M. Bank, J. Leffingwell, and C. Thies, Macromolecules,
    4, 43 (1971).
6. M. Bank, J. Leffingwell, and C. Thies, J. Polym. Sci.
    A-2, 10, 1097 (1962).
7. T. K. Kwei, T. Nishi, and R. F. Roberts, Macromolecules,
    7, 667 (1974).
8. T. Nishi and T. K. Kwei, Polymer, 16, 285 (1975).
9. T. Nishi, T. T. Wang, and T. K. Kwei, Macromolecules,
    8, 227 (1975).
10. R. L. Scott, J. Chem. Phys. 17, 279 (1949).
11. J. W. Cahn, J. Chem. Phys., 42, 93 (1965); Trans. AIME,
    242, 166 (1968).
12. T. Nishi and T. T. Wang, Macromolecules, 8, 909(1975).
13. T. K. Kwei, G. D. Patterson, and T. T. Wang,
    Macromolecules, in press.
14. T. K. Kwei, H. L. Frisch, W. Radigan, and S. Vogel,
    Macromolecules, in press.
15. T. Nishi, T. K. Kwei, and T. T. Wang, J. Appl. Phys.
    46, 4157 (1975).
16. T. Nishi and T. K. Kwei, J. Appl. Polym. Sci.,
    20, 1331 (1976).

DISCUSSION

T. W. HUSEBY - BELL LABORATORIES. My question is in two parts. They both refer to the spinodal decomposition discussion that we have had. I want to draw attention to the fact that systems in metallurgy which spinodally decompose are usually connected with ideas of crystallization, particularly homogeneous nucleation and growth. I think what you were suggesting is that spinodal decomposition does not necessarily have to be associated with homogeneous crystallization. Is that correct?

T. K. KWEI - BELL LABORATORIES. Let me first make a remark that the idea of spinodal decomposition at least in the experimental sense has not been universally accepted even by metallurgists. This is one of the few attempts applied to polymer systems. Since our two polymers do not crystallize and if our conclusions are indeed correct then I will say that spinodal decomposition can occur without crystallization.

T. W. HUSEBY. My second comment refers to the fact that in the metal systems one of the strongest impetuses for utilizing spinodally decomposing systems is their dramatic increases in tensile strength. At least it affords the opportunity for heavy deformation processing that doubles the tensile strength. Do you have any evidence for that in the systems you talk about here? T. K. KWEI. No, we did not measure it. The only thing I can say is that I hope in the PVC-Hycar system what we observe as a result of heat treatment may have something to do with this inter-connectivity.

R. NATARAJAN - LORD CORP. Have you measured the spinodal wavelengths independent of the microscopic studies you showed in PS/PVME systems? T. K. KWEI. No, I have not.

R. NATARAJAN. You did show some data on dynamic, mechanical properties. Have you tried to correlate the spinodal wavelengths with the dynamic, mechanical properties you get? I believe Cahn and Hilliard did this kind of correlation in the case of metallic systems.

T. K. KWEI. In this particular system, polystyrene and PVME with respect to lower critical solution temperatures,

the phenomena is reversible. Once we cool it down to room
temperature it is again compatible. So we cannot carry out
that experiment easily.

A. R. SHULTZ - GENERAL ELECTRIC CO. Do I understand
you correctly that in the possible spinodal decomposition
in your polymeric system the developed domain size is large?
I think that in metallurgical systems in which spinodal
decompositions are attempted the advantage appears to be
in the very fine (small) domains produced. The tensile
improvement in such metals may be due to a lack of serious
defects associated with large grain boundaries. But I
don't think that one would infer from your results that
such an improvement in polymeric systems would be achieved
by spinodal phase separation.   T. K. KWEI. Because I
worked with glass reinforced composite materials, I thought
that interconnecting structures would provide a method of
obtaining in situ reinforcement. But, I really don't know
if that's possible in practice.

A. POCIUS - 3M CO. What range of weight fraction of
polyvinylidene fluoride is necessary to compatibilize the
polymethacrylate polymers?   T. K. KWEI. Roughly 35-40%.
I neglected to mention that too little of the PVF_2 is not
sufficient to bring the two methacrylate polymers together.

R. MOORE - EASTMAN KODAK COMPANY. Your last slide
indicated that the impact strength went through a maximum.
Is this a function of the change in domain size?
T. K. KWEI. We do not know because the morphology was
very difficult to see and, furthermore, after letting the
poly(vinyl chloride) stand exposed to a temperature of
130°C for a long time it does degrade.

R. MOORE. In the case of the OsO_4 staining of the
methacrylates, what is the mechanism that gives rise to the
staining?   T. K. KWEI. There are several possibilities.
One is there might be residual solvent inadvertently left in
one of the phases. Now since PMMA has a high T_g there might
be a small amount of residual solvent left in the PMMA phase
and that may be enough to react with OsO_4. The second
possibility is that somehow the interface is weak and able
to accumulate th OsO_4.

SOME STUDIES ON THE RATES OF CONFORMATIONAL TRANSITIONS AND OF CIS-TRANS ISOMERIZATIONS IN FLEXIBLE POLYMER CHAINS

Herbert Morawetz

Polytechnic Institure of New York
333 Jay Street
Brooklyn, N. Y.   11201

INTRODUCTION

While the conformational distribution in flexible chain molecules has been a subject of intensive theoretical and experimental study ever since Kuhn's pioneering work[1], the dynamics of conformational transition have become only much more recently a subject of experimental study. A variety of techniques have been utilized for this purpose based on dielectric dispersion[2-4], depolarization of fluorescence[5-8], the frequency dependence of sound absorption[9-13], NMR relaxation[14-17] and motional narrowing of the ESR spectra of spin-labelled polymers.[18]

The consistency of these various methods may be evaluated from the results listed in Table I of data obtained with polystyrene in solvents of low viscosity by the sound absorption (SA) and NMR relaxation methods, for poly(p-chlorostyrene) by dielectric dispersion (DD), for polystyrene labelled with nitroxyl by the ESR method and for a styrene copolymer with a small concentration of 9-p-vinylphenyl-10-phenylanthracene residues by depolarization of fluorescence (DF). It may be seen that sound absorption and dielectric dispersion yield similar transition frequencies in the range of 12-35 MHz at 10-25°C,

TABLE I

Literature data concerning conformational
transition rates of polystyrene in dilute
solution

Solvent	Temp. (°C)	Method	Frequency (MHz)	Rel. Time (ns)	Ref.
Toluene	10	SA	3.4;22		12
DMF	25	"	5.8;35		11
Benzene	25	DD	30		3
Benzene	25	"	24		4
Dioxane	23	"	12		4
Ethyl acetate	35	DF		1.2	6
"	R.T.	"		2.1	8
C_2Cl_4	44	NMR		0.6	16
Toluene	44	ESR		0.4	18

although the ultrasonic method also reveals a slower process. By contrast, the rotational relaxation times obtained by NMR, ESR and depolarization of fluorescence lie in the range of 0.4-2 nanoseconds, corresponding to a much faster process.

In this context it is of special interest to enquire what is the relation between the rate of hindered rotation around a specific bond in the backbone of a long chain polymer and a similar process in an analogous low molecular weight compound. Among the experimental techniques listed above, sound absorption is the only one which is, in principle, suitable for providing information on this point, since in all the other techniques results obtained with small molecules will reflect both effects due to conformational transitions and the rotation of the molecule as a whole. With the use of the sound absorption technique, Cochran et al[13] found that the relaxation frequency which they assigned to conformational transitions in the polystyrene backbone first decreases with increasing molecular weight but becomes molecular weight independent for long chains; they fitted their data by a model in which the hindered rotation is 2.5 times as rapid in the terminal ten monomer residues as in the interior of the chain. It seems a pity that no analogous measurements were carried out on well characterized low molecular weight model compounds such as the isomers of 2,4-diphenylpentane or 2,4,6-triphenylheptane.

The belief is widely held that conformational transitions in the backbones of long polymer chains must involve two correlated hindered rotations (a "crankshaft-like motion") so as to avoid the need for a long chain segment to swing through the viscous solvent medium. (See Figure 1). This concept has been used as a basis of theoretical studies of segmental diffusion of polymer chains in dilute solution [19-21] and in the interpretation of experimental data[2,4,9,22].

A   B

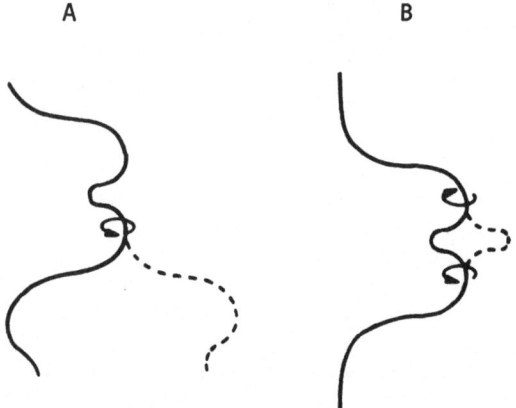

Figure 1.   Schematic representation of conformational transition in a flexible chain molecule:
(a) rotation around a single bond; (b) correlated
rotation around two bonds.

    If this model represents correctly the
physical reality, the free energy of activation
should be substantially higher for hindered
rotation in the backbone of a long polymer chain
than for a similar process in a small molecule,
resulting in a large decrease in the rate of
conformational transitions. The studies described below (in which cis-trans isomerization was
regarded as a special case of conformational
transition) were designed to establish whether
polymers and their analogs exhibit this difference in their behavior.

HINDERED ROTATION AROUND AMIDE BONDS
IN POLYAMIDES AND THEIR ANALOGS.

    In symmetrically disubstituted amides, the
rotation around the amide bond can be studied
conveniently by NMR spectroscopy, since the two

substituents are in different magnetic environ-
ments. Thus, dimethylformamide exhibits two
methyl peaks at low temperature, when the difference between the two chemical shifts $\delta \nu$ is large
compared to the rate constant for hindered rotation k_t, but the two peaks collapse into a
singlet at the temperature at which
$k_t = (\sqrt{2}\pi/2)\delta \nu$. Gutowsky and his collaborators
have developed the theory which yields the
activation parameters for hindered rotation from
the temperature dependence of the shape of the
absorption band[23,24] and we have used this technique for a comparison of the rates of hindered
rotation in piperazine polyamides and their low
molecular weight analogs.[25] Contrary to expecta-
tion, no significant difference was found between the rates of hindered rotation in poly-
(piperazine succinate) and diacetylpiperazine or
poly(piperazine terephthalate) and dibenzoyl-
piperazine.

THERMAL CIS-TRANS ISOMERIZATION OF AZO-
BENZENE RESIDUES IN THE BACKBONE OF
POLYMER CHAINS

Azobenzene derivatives may be photoisomeriz-
ed from the stable trans form to the cis isomer
and the reverse reaction takes place at a convenient rate by a thermal mechanism in the
neighborhood of ambient temperatures. Since the
isomerization is accompanied by a large shift in
the ultraviolet spectrum, the kinetics of both
the photochemical and the thermal process may be
followed by UV spectroscopy. We have utilized
this technique for comparing rate constants of
the thermal reaction[26,27] and the photochemical
quantum yields[28] in polymers containing azobenzene residues and in their low molecular
weight analogs.

The use of UV spectroscopy has two important
advantages compared to the NMR method for the
study of rates of conformational transitions:
(1) The NMR technique is limited to media of low

viscosity which yield spectra of high resolution.
No such limitation exists when UV spectroscopy
is the analytical tool and we may study by this
method systems ranging from dilute polymer solutions to rigid polymer glasses. (2) While the
NMR method yields only an average rate constant,
the UV analysis allows us to follow precisely
the progress of the isomerization process.
Thus, any deviation from first-order kinetics
will indicate that the azobenzene residues in
the system under study isomerize at different
rates. Such a dispersion of the rate constant
for hindered rotation was postulated by other investigators. Cochran et al[13] reached this conclusion on the basis of the molecular weight dependence of the relaxation times observed with
polystyrene by sound absorption studies. Valeur
and Monnerie[8] subjected a solution of polystyrene
with anthracene residues incorporated into the
chain backbone to a nanosecond light flash and
found that the time-dependence of the anthracene
fluorescence deviated from a simple exponential
decay; this was ascribed to differences in the
local conformational mobility of the chain.

A study of the dark reaction following a
photochemical trans-cis isomerization of dilute
solutions of polyamides containing small concentrations of azobenzene residues in the chain
backbone[26] yielded the following results: (1) The
reaction was strictly first-order. (2) There was
no significant difference between the rate constant characterizing azobenzene residues in the
polymer chain backbone and analogous low molecular
weight azobenzene derivatives. (3) The rate
constant of the cis-trans isomerization was not
reduced at a polymer concentration corresponding
to extensive interpenetration of the flexible
chain molecules.

THERMAL CIS-TRANS ISOMERIZATION OF AZO-
BENZENE RESIDUES ATTACHED TO POLYMERS IN
THE GLASSY STATE

When films of glassy copolymers of methyl
methacrylate with a small proportion of p-(N-

methacrylyl)aminoazobenzene are irradiated to effect <u>trans-cis</u> isomerization and the light is switched off, the dark reaction exhibits a strong deviation from first-order kinetics.[27] This phenomenon is characteristic of the glassy state, since the dark reaction follows first-order kinetics if the film is converted to the rubbery state by incorporation of a plasticizer (Figure 2). The thermal reaction in the plasticized polymer (or, in the polymer in bulk above the glass transition temperature) takes place at a rate indistinguishable from that observed in

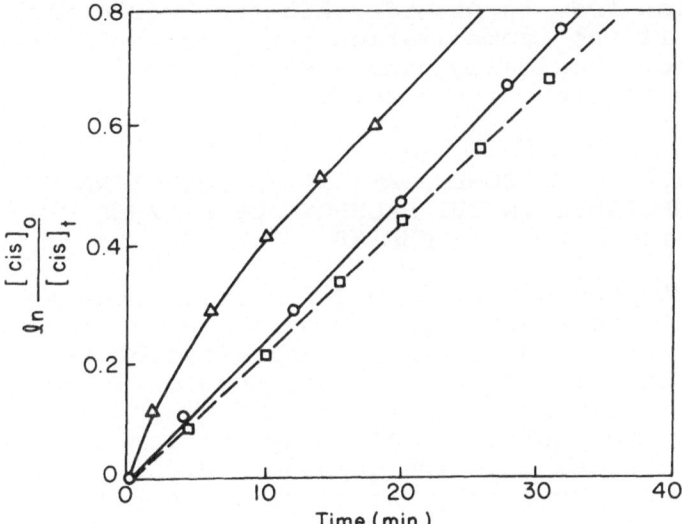

Figure 2. Thermal <u>cis-trans</u> isomerization of methyl methacrylate copolymer with 0.9 mole % of p-(N-methacrylyl)aminoazobenzene at 60°C after photochemical <u>trans-cis</u> isomerization at the same temperature: ($\triangle$)bulk polymer, (O)polymer plasticized with 30% dioctyl phthalate,($\square$)dilute solution in butyl acetate.

dilute solution. In the glassy polymer, some of
the azobenzene residues react at a rate which is
anomalously fast. This anomaly is related to the
fact that the preceding photochemical process was
carried out on a glassy material- if the photochemical trans-cis isomerization is carried out
above Tg and the sample is cooled through Tg under irradiation, the subsequent dark process
follows first-order kinetics.

The dispersion of the rate constant characterizing chemically identical groups may be
thought of as a consequence of the non-equilibrium
nature of the glassy state which corresponds to a
fluctuation of the local free volume. Similar
phenomena were observed with isomerizing small
molecules incorporated into a glassy polymer
matrix.[29] Recently, it has been suggested that a
deviation from first-order kinetics would also
result if the isomerization requires the diffusion
of a photochemically generated conformational defect to the reaction site.[30]

THE PHOTOISOMERIZATION OF AZOBENZENE
RESIDUES IN THE BACKBONE OF POLYMER
CHAINS

Both the hindered rotation around the amide
bond and the thermal cis-trans isomerization of
azobenzene residues are characterized by high
energy barriers in the neighborhood of 20 kcal/
mole. Some years ago, Malkin and Fischer[31]
studied the photochemical isomerization of azobenzene and found that the quantum yield is temper-
ature-dependent; they interpreted their data as
reflecting an energy barrier of 2-3 kcal/mole between the excited cis and trans species. A comparison of the quantum yields for the photoisomerization of azobenzene residues in the backbone of
polymer chains and in low molecular weight analogs
should, therefore, indicate whether conformational
transitions characterized by low activation

energies are significantly more difficult in the backbones of polymers than in small molecules.

The experimental results revealed no significant difference in the rate of photoisomerization of azobenzene residues in the backbone of polyamides and in low molecular weight analogous azobenzene derivatives when both were studied in dilute solution.[28] However, while the photochemical reactivity of the small species was relatively insensitive to the concentration of added polymer, the quantum yield for the photoisomerization of the azobenzene residues in the polymer backbone dropped precipitously with increasing concentration. In a glassy polymer film containing 8% DMSO plasticizer, the quantum yield for the isomerization of the polymer was reduced by a factor of 2500 while it was reduced only by a factor of 5 for the small molecule (Figure 3).

STUDIES OF RATES OF CONFORMATIONAL TRANSITIONS BY EXCIMER FLUORESCENCE

The emission spectra of certain aromatic molecules exhibit a characteristic change in concentrated systems which is due to the emission of a sandwich complex of an excited molecule with a second molecule in the ground state. Such complexes are called excimers.[32] In 1965, it was reported that dilute solutions of α,ω-diphenyl-alkanes, $\phi-(CH_2)_n-\phi$, have an emission spectrum similar to that of toluene, except for the compound with n=3 which was characterized by a typical excimer fluorescence.[33] It is certain that a conformation of 1,3-diphenylpropane in which the phenyl groups lie parallel to each other would have a prohibitive energy in the ground state and it must then be assumed that the conformation required for excimer formation (which is strongly exothermic) requires a hindered rotation during the lifetime τ of the excited state:

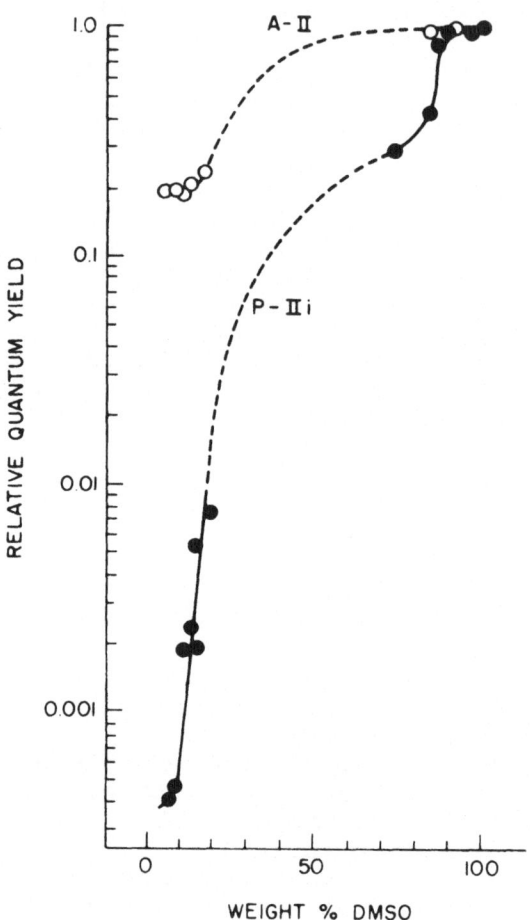

Figure 3. Relative quantum yield of <u>trans-cis</u> photoisomerization of polymer containing azobenzene residues in its chain backbone and of its low molecular weight analog in mixtures of poly (isophorone diamine isophthalamide) and dimethylsulfoxide. Polymer (P-IIi): Copolyamide of isophthalic acid with isophorone diamine containing 1.5 mole % 4,4'-azodianiline. Analog (A-II): 4,4'-(dibenzoyl)-azodianiline.

Thus, the relative fluorescence yield from the monomer and the excimer provides, in principle, information about $k\tau$ and since τ is of the order of 10^{-8} sec, this technique is suitable for the study of hindered rotations with rather low energy barriers.

The study of intramolecular excimer formation in polymers leads to some interesting insights. The emission spectra of polystyrene[34] and its dimer analog 2,4-diphenylpentane[35] are remarkably similar, suggesting once again that the rates of conformational transition must be nearly the same in a polymer backbone and an analogous small molecule. A surprising result was obtained in a study of the emission spectra of acenaphthylene polymers and copolymers.[36] Steric restrictions in the homopolymer are too severe to allow excimer formation from neighboring monomer residues and it had therefore been suggested[37] that excimers must form, in this case, from next-to-nearest monomer residues. Our data[36] support this interpretation, since the ratio of normal and excimer emission is identical in polyacenaphthylene(I) and the alternating acenaphthylene-maleic anhydride copolymer (II)(It should be noted that no

excimer emission is observed from the alternating styrene-maleic anhydride copolymer). However, in

the four alternating copolymers of acenaphthylene
III a,b,c,d,

III   (a) X=H, Y=COOCH$_3$
       (b) X=H, Y=CN
       (c) X=CH$_3$,Y= COOCH$_3$
       (d) X=CH$_3$ Y=CN

excimer emission was found to be critically dependent on the presence of a quaternary carbon in
the comonomer, being very strong in IIIc and
IIId and almost absent in IIIa and IIIb. This result seems to indicate the high sensitivity of
excimer emission to structural features affecting the conformational mobility of polymer chains.

Intramolecular excimer emission is also a
valuable tool for the study of the dependence of
the rate constant for a conformational transition
k_t on the properties of the solvent medium. This
rate constant is related to the concept of the
"internal viscosity" η_i of a polymer chain which
opposes the separation of the chain ends at a
rate dh/dt in response to a force F so that dh/dt=
F/η_i. Kuhn and Kuhn who originally introduced the
concept[38] assumed that the internal viscosity was
a function of the potential energy barriers
characterizing hindered rotation around the bonds
of the chain backbone; it was, therefore, assumed
to be an intrinsic property of the chain, independent of the solvent medium. More recently, it
has been pointed out that the viscosity η of the
medium must necessarily contribute to the resistance to conformational changes, so that η_i should
be the sum of two contributions, one due to the
height of the potential energy barriers, the
other proportional to the viscosity of the solvent, i.e., η_i= A+Bη.[39] A similar reasoning[6]
would lead us to expect k_t to be of the form
k_t= $\underline{k}$T/(A'+B'η) and studies of the dependence of

the ratio of normal and excimer fluorescence of dibenzyl ether on the viscosity of the solvent seem to substantiate this analysis.[40] It is, of course, possible that specific solvation effects also affect the magnitude of k_t and this is being currently investigated in our laboratory on a series of model compounds.

DISCUSSION

The most important issue in this series of investigations was the question whether two hindered rotations have to be correlated in time to produce a conformational transition in the backbone of long polymer chains in dilute solution. It seems to me that the evidence against this concept is overwhelming, since transitions in polymers and their low molecular weight analogs were found to take place at essentially identical rates. This is the case also for transitions characterized by quite low energy barriers as ex-emplified by the data on photochemical isomeriza-tion of azobenzene residues in the backbone of polymer chains and the conformational transitions required to produce excimer fluorescence in poly-styrene. Yet, as pointed out earlier, it is in-conceivable to have a single conformational transition result in a rapid motion of a long sec-tion of the chain through the viscous medium as represented schematically in Figure 1a. This apparent contradiction may be resolved if we assume that the hindered rotation does not take place in a single kinetic step, but that it in-volves a large number of small oscillations of the internal angle of rotation, with all the re-sulting energy states in equilibrium with each other. The transition state will then also be in equilibrium with the ground state and since the height of the barrier to internal rotation is the same in the polymer and its analog, the rate of hindered rotation will also be identical. This interpretation may be considered to represent an application of Kramers' theory of the diffusion of a particle across an energy barrier[41].

The dependence of the conformational mobility
on the polymer concentration depends critically
on the height of the energy barrier. When this
barrier is high, the arguments of the Kramers
theory hold up to high polymer concentrations,so
that hindered rotation (or cis-trans isomerization) is not slowed down by the mutual entanglement of polymer chains.[26] By contrast, in a
process in which a change in the shape of the
chain involves only a low energy barrier, the
process is strongly impeded in concentration systems.[28] This may be interpreted either as evidence
for the cooperativity of transitions taking place
in neighboring chains or as evidence for the re-
quirement of a "crankshaft-like motion" in any
one chain. We may recall that the concept of such
a motion was advanced by Schatzki specifically
for the interpretation of a behavior of polymers
in bulk.[42]

Acknowledgement:- I am indebted to the National
Science Foundation for financial support of this
study by Grant DMR 75-05234.

REFERENCES

1.   W. Kuhn, Kolloid-Z., 68, 2 (1934).
2.   W. H. Stockmayer, Pure Appl. Chem., 15, 539
      (1967).
3.   W. H. Stockmayer and K. Matsuo, Macromole-
      cules, 5, 766 (1972).
4.   S. Mashimo, Macromolecules, 9, 91 (1976).
5.   E. V. Anufrieva, M. V. Volkenstein,
      M. G. Krakovyak and T. V. Sheveleva, Dokl.
      Akad. Nauk SSSR, 186, 854 (1969).
6.   D. Biddle and T. Nordström, Ark. Kemi, 32,
      359 (1970).
7.   L. Monnerie and S. Gorin, J. Chim. Phys.,
      67, 400 (1970).
8.   B. Valeur and L. Monnerie, J. Polym. Sci.,
      14, 11, 29 (1976).
9.   O. Fünfschilling, P. Lemarechal and R. Cerf,
      Chem.Phys.Lett., 12, 365 (1971).

10. W. Ludlow, E. Wyn-Jones and J. Rassing,Chem. Phys.Lett., 13, 477 (1972.

11. P. Lemarechal, Chem. Phys. Lett., 16, 495 (1972).

12. B. Froelich, C. Noel, J. Lewiner and L. Monnerie,Comp.Rend.,Ser.C,277,1089(1973).

13. M.A. Cochran, J.H. Dunbar, A.M.North and R.A. Pathrick, J.Chem.Soc.,Faraday II, 70,

14. K. J. Liu and R.Ullman, J.Chem.Phys.,48, 1158 (1968).

15. K.J. Liu and J.E.Anderson, Macromolecules, 3, 163 (1970).

16. A. Allerhand and R.K.Hartstone, J.Chem.Phys., 56, 3718 (1972).

17. G. Hermann and G.Weill,Macromolecules, 8, 171 (1975).

18. A.T. Bullock, G.G. Cameron and P. M. Smith, J.Phys.Chem., 77, 1435 (1973).

19. P.H. Verdier and W.H. Stockmayer, J. Chem. Phys., 36, 227 (1962).

20. L. Monnerie and F. Geny, J. Polymer Sci., Pt.C, 30, 93 (1970).

21. E. Helfand, J. Chem. Phys., 54,4651 (1971).

22. W. Pechhold, S. Blasenbrey and S. Woerner, Kolloid-Z.-Z.Polymere, 189, 4 (1963).

23. H.S. Gutowsky, D.W.McCall and C.P. Slichter,Jr., Ric.Sci.,21, 279 (1953).

24. H.S. Gutowsky and A.Saika, Ric. Sci., 21, 1688 (1953).

25. Y.Miron, B.R.McGarvey and H. Morawetz, Macromolecules, 2, 154 (1969).

26. D.Tabak and H.Morawetz, Macromolecules, 3, 403 (1970).

27. C.S. Paik and H.Morawetz, Macromolecules, 5, 171 (1972).

28. D. T.-L.Chen and H.Morawetz, Macromolecules, 9, 463 (1976).

29. W.J. Priest and M.M.Sifain, J.Polymer Sci., Part A-1,9, 3161 (1971).

30. M.Kryszewski and B. Nadolski, Paper presented at the Microsymposium on Photochemical Processes in High Polymers, Leuven,Belgium, June 2-4, 1976.

31. S. Malkin and E. Fischer, J.Phys.Chem., 66, 2482 (1962).

32. J.B. Birks, Prog.React.Kin., 5,181 (1970).
33. F. Hirayama, J. Chem. Phys., 42,3163 (1965).
34. J.W. Longworth, Biopolymers, 4,1131 (1966).
35. J.W.Longworth and F.A. Bovey, Biopolymers, 4,
     (1966).
36. Y.-C. Wang and H.Morawetz, Makromol.Chem.,
     Suppl. 1, 283 (1975).
37. C. David, M. Lempereur and G. Geuskens,
     Eur.Polym.J., 8, 417 (1972).
38. W. Kuhn and H. Kuhn, Helv.Chim.Acta, 29,
     609,830 (1946).
39. A. Peterlin, J.Polym.Sci., B 10, 101 (1972).
40. Y.-C. Wang and H. Morawetz, J. Am. Chem. Soc.,
     98, 3611 (1976).
41. H. A. Kramers, Physica, 7, 284 (1940).
42. T. Schatzki, Polym.Prepr., Amer.Chem.Soc.,
     Div.Polym.Chem., 6, 646 (1965).

DISCUSSION

C. OVERBERGER - U. OF MICHIGAN. I was not completely clear on the methyl methacrylate polymer with the azo side chains. When you irradiate that material below the glass transition, cut off the light and then watch the dark reaction, do all azo linkages return to the trans state? How long does that take, approximately?   H. MORAWETZ - POLYTECHNIC I. OF NEW YORK. This is a process which has a half-life of 10 minutes at 70°C. There is one thing I didn't say, that is, that in the rubbery state there is absolutely no difference in the thermal reaction rate of the rubber and the dilute solution.

C. OVERBERGER. In acenaphthylene copolymers, where you were getting excimer formation not by nearest neighbor interaction but by next-to-nearest neighbor interaction, there's a big difference between the methyl acrylate and methyl methacrylate copolymers. Is there some reason to expect that this is due to differences in obtaining the aromatic groups near each other?   H. MORAWETZ. No, we did not expect it. The most probable conformation will be, of course, different in the case of the copolymer containing quarternary carbons. We know it is different because the NMR peaks of the aromatic residues are very sensitive to shielding by the next aromatic residue and we can see that the aromatic region of the NMR spectrum is very, very different in the methyl acrylate and the methyl methacrylate copolymer.

C. OVERBERGER. So it might really be that the interaction responsible for excimer formation is not 1,3; it might be 1,5.   H. MORAWETZ. I doubt it very much. There are some recent studies by Zachariasse in Göttingen [Zachariasse and Kühnle, Z. phys. Chem., N.F., 103, 267 (1976)] who showed that if you have pyrene residues where the external interaction is very, very much more powerful, then the condition that the two residues have to be parallel to each other is very much relaxed. In that case you get very significant excimer formation even if your two pyrenyl residues are quite far from one another but with phenyl or naphthyl residues this has never been observed. The indication is that if they are further apart than 3 or maybe 4 carbon atoms from each other, then excimer formation is negligible.

F. H. WINSLOW - BELL LABORATORIES. I wasn't sure I
understood what you were saying in your last slide. For
example, was it primarily a polar effect or a conformational
effect that you observed? It is known that polar solvents
have a tremendous effect on excimer formation. Some of the
polymers had rather interesting polar groups.

H. MORAWETZ. It is a conformational effect. All the
materials in the last slide were in the same solvent.
That is not what we varied. We varied the nature of the
polymer. As to polar groups, we found that methyl acrylate
and acrylonitrile were identical in their behavior but
methyl methacrylate and methacrylonitrile were completely
different and so this, in my opinion, was clearly a
reflection of the presence of the quaternary carbon atom.
Now, of course, we cannot say for sure whether this is due
to a difference in the conformational distribution of the
ground state or a difference in the rate of conformational
transition. We know that even the ground state distribution
is different but very often people think that just because
poly(methyl methacrylate) is very crowded that this means
that the potential energy barriers must be very high for
conformational transitions. That doesn't follow. All
conformations in poly(methyl methacrylate) represent states
of very high potential energy. The barriers between them
may not be terribly high for all we know.

TOPOCHEMICAL EFFECTS IN CHEMICAL REACTIONS OF CRYSTALLINE

ORGANIC COMPOUNDS

Jerome B. Lando

Department of Macromolecular Science
Case Western Reserve University
University Circle
Cleveland, Ohio 44106

INTRODUCTION

A topochemical effect in a solid state reaction is any effect on the structure and properties of the product or the kinetics of the reaction that can be directly attributed to the geometric arrangement of the reacting groups or the distance between those groups. Such effects can conveniently be classified in three catagories: (a) Reactions in which the reactant and product form solid solutions over the entire range of compositions. (b) Reactions in which the product phase is nucleated but a crystallographic relationship between the reactant and product phases is maintained. (c) Reactions which result in an amorphous product or crystalline product having no crystallographic correlation with the reactant.

The following presentation is not intended to be a review of this field in the usual sense. Rather, an example of each class of reactions will be discussed: (a) The polymerization of o,o-diacetylenediphenyl glutarate and 5,7-dodecadiinediol-1,12-bis-phenylurethane. (b) The cyclization reaction of N-(para-chlorophenyl) phthalanilic acid to N-(para-chlorophenyl) phthalimidle with the elimination of water. (c) The polymerization of crystalline methacrylic acid. Finally a discussion will be undertaken of the polymerization of molecular multilayers of a number of monomers, such as vinyl stearates and α-octadecylacrylic acid, which may not fit conveniently into any of the above categories.

189

A central theme will be an examination of the molecular
mechanism of these reactions and how the structure and
properties of the product is related to reactant structure
and morphology.

DIACETYLENE POLYMERIZATION

Many substituted diacetylenes have been found to be
highly reactive in the solid state(1). By exposure to uv,
x-ray or gamma radiation or by annealing topochemical polymerization is initiated. Each monomer molecule joins with
two neighboring molecules in a 1,4-addition reaction at
the conjugated triple bonds to form a linear fully conjugated
polymer chain. In this way large, nearly defect-free
polymer single crystals can be obtained(1). During such
a reaction unit cell dimensions show only slight changes,
the space group is retained and a solid solution exists
at all conversions(1). From x-ray diffraction(2,3) and
Raman spectral studies(4) the polymer backbone is best
represented by the mesomeric structure I (acetylenic)

$(R=-H_2C-O-\overset{\text{O}}{\underset{\text{||}}{C}}-NH-\langle\text{O}\rangle,(2)\,;\ R=-H_2C-O-\overset{\text{O}}{\underset{\text{||}}{S}}-\langle\text{O}\rangle-CH_3,(3)\,.$

However, there is in some polymer spectroscopic evidence
for significant resonance contribution from structure II
(butatriene) corresponding to considerable π-electron
delocalization(4).

In contrast to the crystal structures(2,3), in which

primary interactions between the side groups are intermolecular, we have undertaken the structure determination
of two polydiacetylenes in which primary side group interactions are intramolecular:

1) poly(5,7-dodecadiinediol-1,12-bis-phenylurethane)(5)

$R=-(CH_2)_4-O-\overset{\text{O}}{\underset{}{C}}-\overset{H}{\underset{}{N}}-$ ⬡ (interaction through hydrogen
bond)

and

2) poly(o,o-diacetylenediphenyl glutarate)(6)

$2R=$ ⬡ $-O-\overset{O}{\underset{}{C}}-(CH_2)_3-\overset{O}{\underset{}{C}}-O-$ ⬡ (interaction through a
covalent bond).

It is expected that strong intramolecular side group
interactions as opposed to intermolecular interactions
might effect the electronic structure of the polymer backbones. In addition the existance of a stable solid solution
of o,o-diacetylenediphenyl glutarate and the corresponding
polymer indicated the possibility that a complete crystal
structure would yield detailed information concerning the
molecular motions necessary for this type of topochemical
polymerization.

Intensity data were collected using a Picker Facs-I
computer controlled four-circle single crystal diffractometer. Unit cell data for poly(5,7-dodecadiinediol-1,12-
bis-phenylurethane) are $\underline{a}$=6.229Å, $\underline{b}$=39.03Å, $\underline{c}$=4.909(chain
axis), β=106.85 degrees, space group $P2_1/c$, and Z=4. The
observed intensity data consisted of 724 independent
reflections.

Unit cell data for the solid solution of o,o-
diacetylenediphenyl glutarate and its polymer are $\underline{a}$=23.12Å,
$\underline{b}$=7.87Å, $\underline{c}$=9.69Å(chain axis), β=111.29 degrees, space
group C2/c, and Z=4. The intensity data consisted of 902
independent reflections.

The bc projection of poly(5,7-dodecadiinediol-1,12-
bis-phenylurethane) is shown in figure 1. Although not
indicated in this figure reasonable anisotropic temperature
factors were obtained for all atoms. Hydrogen atom temperature factors were held isotropic. The final residual R=0.08.
Standard deviations for bond angles were 1 degree while

standard deviations for bond lengths were 0.01-0.02Å.   The
polymer backbone is planar within these limits.

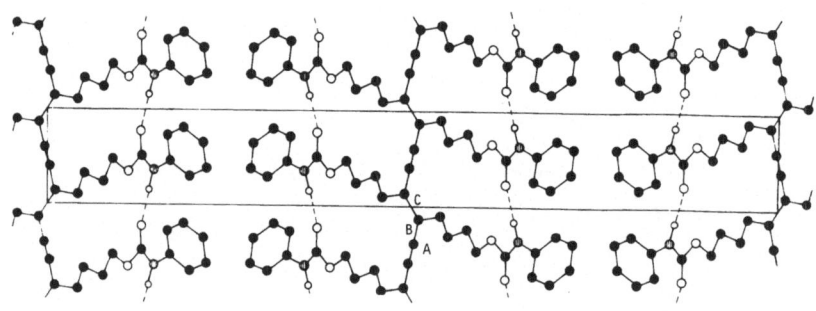

Figure 1.   Projection of Poly(5,7-Dodecadiinediol-1,12-
bis-Phenylurethane) on the bc Plane.

The ac projection of one polymer chain and the
corresponding monomer stack of o,o-diacetylenediphenyl
glutuate are shown in figure 2b and 2a respectively.   Thermal
ellipsoids and hydrogen atoms are indicated in this figure.
Only four atoms in the assymetric unit of structure (half
a monomer unit) had sufficiently different positions (over
0.4Å) to be assigned separate positions in the monomer and
polymer structures.   These were the two atoms of the
diacetylene rod and two of the phenyl group carbons (the
carbon bonded to the rod and the carbon para to it).   Of
these four atoms only the atom between bonds B and C were
sufficiently far apart to refine anisotropic temperature
factors.   The other three carbon atoms along with all
hydrogen atoms were held isotropic.   The large anisotropic
thermal elipsoids of three of the remaining four carbon
atoms reflect their unresolvable movement upon polymerization.
The remainder of the molecule had almost identical positions
in the monomer and polymer indicated by normal thermal
ellipsoids.   Standard deviations of bond angles in this
structure were around 1 degree, while standard deviations
of bond lengths for the monomer were from 0.01-0.02Å and
for the polymer 0.01-0.03Å.   On the basis of infrared spectra

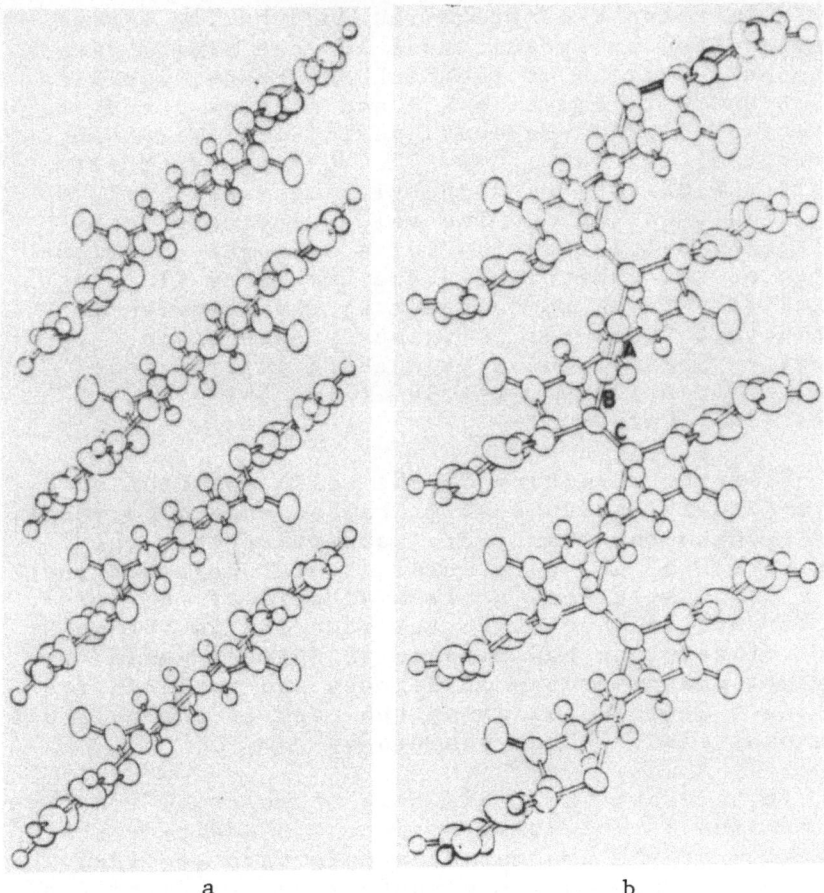

a   b

Figure 2.   Projection of One Polymer(b) and the
Corresponding Monomer Stack(a) of o,o-Diacetylenediphenyl
Glutarate on the ac Plane.

on the one hand and the minimization of the residual
(R=0.075) and the standard deviations on the other hand,
the solid solution was found to contain 35% polymer and
65% monomer. Ring strain causes the diacetylene rod of
the monomer to bend 3° and a nonplanarity of the connecting
bond of the polymer backbone of 3°.

In these two structures (both having strong intramolecular interactions between side chains) the bond lengths of the backbone bonds, labelled A, B and C in equation (1) and figures 1 and 2b, are for poly(5,7-dodecadiinediol-1,12-bis-phenylurethane) A=1.24 Å, B=1.37 Å, C=1.44 Å and for poly(o,o-diacetylenediphenyl glutamate), A=1.29 Å, B=1.38 Å, C=1.42 Å. The values indicate some electron delocalization but a stronger contribution of the butatriene structure (form II in equation (1)) in sharp contrast to the acetylenic structure found for polydiacetylenes having primarily intermolecular side chain interactions A=1.21 Å, B=1.41 Å, C-1.36 Å (2). A=1.19 Å, B=1.43 Å, C=1.36 Å (3).

From the structure of the solid solution of o,o-diacetylenediphenyl glutamate and to polymers. if we make the reasonable assumption that the structure of the pure material are unchanged in the solid solution, a clear picture of the molecular motions necessary for reaction is obtained. The diacetylene rod rotates 33 degrees while the phenyl group rotates 19 degrees and turns in its plane 6 degrees, allowing the rest of the molecule to maintain the same positions.

It is clear that this type of reaction requires a minimum of molecular motion. In addition the details of the monomer structure have considerable influence on the structure and properties of the resultant product.

POLYMERIZATION OF MOLECULAR MULTILAYERS

Molecular multilayers are prepared from monomer molecules that are capable of forming "condensed" monolayers at the air-water interface. These materials are characterized by relatively polar hydrophobic "head" groups and long hydrocarbon "tail" groups. Thus, in a condensed monolayer the hydrocarbon chains tend to pack as do parafins, while the polar groups interact with the water surface.

Multilayers are prepared from monolayers, condensed under lateral pressure, by a technique originally developed by Langmiur and Blodgett (7), which involves dipping a backing plate up and down through the air-water interface. In general, the deposition variables (dipping speed and surface pressure) can be used to obtain deposition of a monolayer on both up and down trip (Y deposition) or either on the up (Z deposition) or down trip (X deposition) alone within limits imposed by the nature of the depositing molecules. X and Y deposition do not necessarily lead to X and Y layers. The facts indicate that, depending on the substance and type of deposition the outermost layer flips over as the multilayers are built, to yield a structure similar to their normal crystal structure (8,9).

In addition, subsequent to building, phase changes occur in many monomer multilayers to yield their normal crystal structures (9,10,11).

Electron and x-ray diffraction studies of the morphology of monomer multilayers of vinyl stearate, (10), α-octadecyl acrylic acid (9), and the cadmium salt of the octadecyl monoester of funaric acid (11) indicate that these multilayers form first with an order-disorder hexagonal packing of the hydrocarbon chains, probably similar to the packing of the condensed monolayer at the air-water interface. Vinyl stearate transforms in a few seconds while α-octadecyl acrylic acid and the fumarate (except the first 10-12 layers) transform in a few hours to hexagonally twinned crystals of their normal crystal structure. Such behavior appears to occur frequently in low molecular weight materials (9). The first 10-12 layers of the fumaric acid derivitive remain in the order-disorder hexagonal structure. Apparently the kinetics of these transformations are related to the strength of interaction of the polar groups.

As has been previously indicated the solid state polymerization of these monomer multilayers does not fit neatly into any of the three catagories of topochemical effects. In all of the multilayers, discussed above, polymerization occurs with solid solution formation (9,10, 11,12). At low conversions there exists solid solutions of polymer side chains in the monomer structure, while at higher conversions there is a phase change to a solid solution of monomer side chains in the order-disorder

hexagonal structure typical of the polymer side chains.
The two phases are not clearly observed simultaniously.
Thus there is an analogy to the previously discussed
polymerization of diacetylene monomers. However, there is
no evidence for stereoregularity in these polymers and
thus backbone participation in these solid solutions. In
fact, no polymer stereoregularity was observed in the
analogous bulk polymerization of crystalline vinyl stearate (13). Thus while the side chains anchor the polymerization, the backbone apparently does not crystallize.

Figure 3 demonstrates both the structure of the
monomer multilayers and the solid solution of side chains
formed during polymerization of Cd-n-octadecylfumarate.
Because of the orientation of these multilayers the powder
diffractometer traces can show only maxima from sets of
planes (00ℓ) parallel to the plate. Thus for 11 Y layers
having a Y structure the order disorder hexagonal monomer
structure is observed and a solid solution of monomer and
polymer side chains with a slight decrease in spacing
with conversion is observed (57Å to 54Å). In the case of
19 layers the monomer is observed to be present in both
its normal monoclinic crystal structure (the phase change
was allowed to occur before these experiments) and in the
order disorder hexagonal form (initial layers). At 40
percent conversion all remaining monomer has converted to
the hexagonal form. Sixty-one layers exhibit the same
effects, although the initial structure is primarily that
of the normal monomer crystal structure.

The possibility of utilizing these multilayer reactions
to produce unique polymer structures is demonstrated in the
production of poly(vinyl stearate) with head-head, tail-
tail (Y) side chain structure (12). In the building of
vinyl stearate multilayers by Y deposition the outermost
layer turns around as the slide passes through the air-
water interface giving X or head-tail structure. Polymerization yields polymer with this type of side chain structure
having a side chain spacing of 28Å (x-ray diffraction)
order-disorder hexagonal packing of the side chains (4.16Å
by electron diffraction). By polymerizating each tail-to-
tail bilayer under the water surface (before the turnaround)
head-head, tail-tail side chain structures are built, which
have a side chain spacing of 56Å (x-ray diffraction) and
orthorhombic side chain packing similar to polyethylene

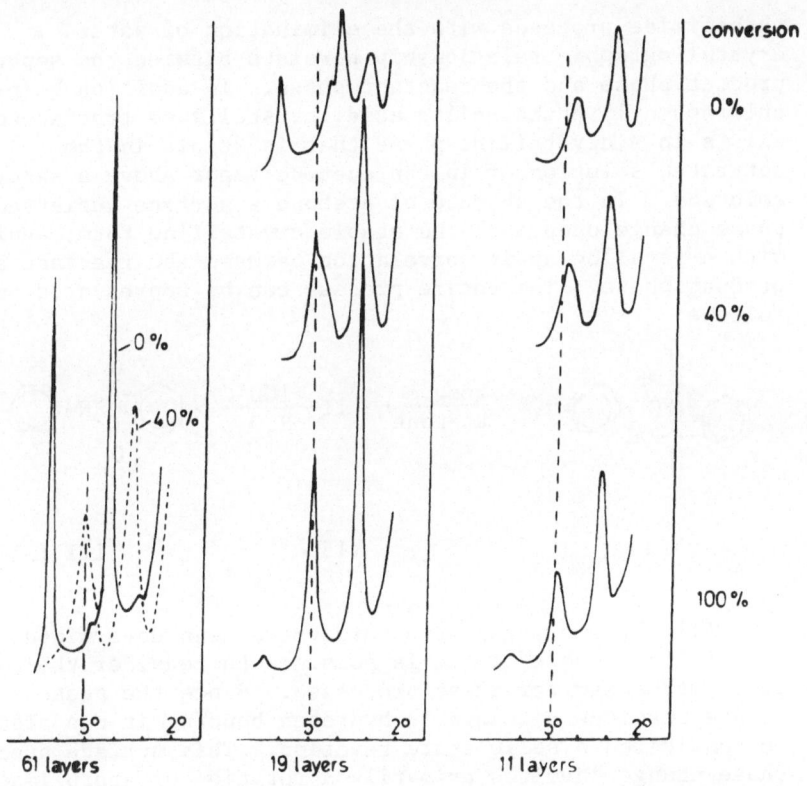

Figure 3.   Diffraction Maxima From 00ℓ Planes During the
Polymerization of Cd-n-Octadecylfumarate Multilayers
(61, 19, and 11 layers with Y structure at 0, 40% and
100% conversion)

(a=7.42Å, b=4.92Å by electron diffraction). There is the
distinct possibility in this unusual polymerization that
syndiotactic polymer is produced in contrast to the normal
multilayer and solid state polymerization of this type of
vinyl monomer (12).

THE CYCLIZATION REACTION OF N-(p-CHLOROPHENYL)
PHTHALANILIC ACID TO N-(p-CHLOROPHENYL) PHTHALIMIDE

Although the cyclization reaction of crystals of N-(p-
chlorophenyl) phthalanilic acid to N-(p-chlorophenyl)

phthalimide proceeds with the elimination of water, a
crystallographic relationship persists between the separate
product phase and the reactant phase. In addition N-(p-
chlorophenyl) phthalanilic acid, crystallized from acetone,
exists in a crystalline phase that is stable in the
saturated solution or in the acetone vapor above a saturated
solution. In the absence of acetone a surface nucleated
phase change occurs to the stable crystalline form, again
with crystallographic correlation between the reactant and
product phase. The entire process can be represented as
follows:

I (14)   (15)   III (16)

     All three crystal structures have been determined
(14,15,16). Therefore it is possible to consider the molecular mechanism for these processes. Since the phase
change involves a change in hydrogen bonding it can also
be considered a solid state reaction. This surface nucleated
phase change involves primarily a rotation of approximately
180 degrees around the indicated bond. The hydrogen bonding changes from dimeric acid hydrogen bonding to polymeric
type hydrogen bonding involving the $\overset{O}{\underset{C-N}{\parallel}}\overset{H}{|}$ group as well as the
acid group. This stable phase has two conformers one
being formed by an additional 30° rotation around the
phenyl-nitrogen bond(14).

     A model for the cyclization reaction is shown in
Figure 4 (16). Only slight rotation around the phenyl-
nitrogen bond and the phenyl-acid carbon bond are required
for reaction. This presumably nucleates at defects in the
polymeric hydrogen bonding. Changes in molecular orientation
are very small while lateral packing of the product imide
(lower molecular volume) creates voids which allow the
diffusion of water out of the crystal in the crystallographic direction parallel to the long dimension of the imide
molecule.

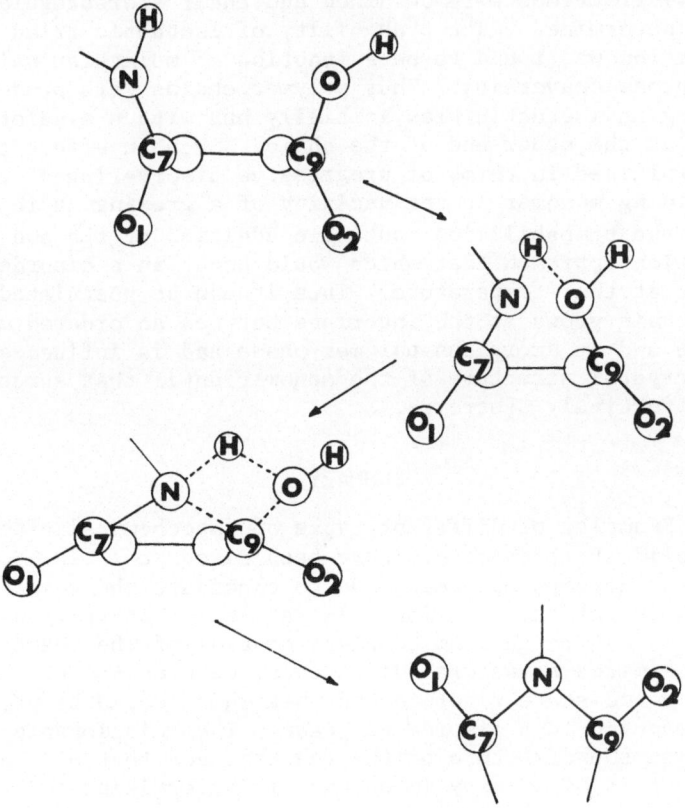

Figure 4. Postulated Molecular Mechanism for the Acid-Imide Conversion

POLYMERIZATION OF CRYSTALLINE METHACRYLIC ACID

Crystalline methacrylic acid can be polymerized at 0°C after ^{60}Co gamma irradiation at -78°C to a dose of .23 Mrad (17). Although the resulting polymer is amorphous the stereoregularity was found to vary with extent of conversion. Since molecular weight was found to be a linear function of conversion the possibility existed that stereoregularity was changing with molecular weight. Although the molecular weight distribution was relatively narrow, a high molecular wt (∿10%) and low molecular weight

($\sim$10%) fractions were obtained and their stereoregularity were determined. The probability of isotactic triad formation was found to be a function of molecular weight not gross conversion. Thus polymer, chains were produced having an isotactic bias initially but with a syndiotactic bias at the other end of the chain. Such an effect can be explained in terms of progressive disordering of the remaining monomer in the vacinity of a growing chain, such that the probabilities isotactic addition at the end of the reaction approach that which would occur in a disordered phase at that temperature. Thus it can be postulated that the chain grows at the interface between an ordered monomer phase and an amorphous polymer phase and is influenced by the crystal structure of the monomer until that structure is effectively destroyed.

SUMMARY

Examples of different types of topochemical effects in solid state reactions have been discussed. In each case an attempt has been made to elucidate the molecular mechanism of the reaction. In is clear that topochemical effects can range from complete control of the reaction to more limited structural differences between the product of a solid state reaction and that which would be produced by reaction in a disordered phase. The relationship between the structure of the reactant and that of the product is of primary importance in determining the degree of topochemical control.

ACKNOWLEDGEMENT

The work of the many people listed in the references is gratefully acknowledged. I would also like to acknowledge the partial support of this work under NATO research grant No. 913, by a NATO postdoctoral research grant to V. Enkelmann, and under NSF grant GK 43432.

REFERENCES

1. G. Wegner, Advances in Chemistry Series, No. 129, 255 (1973).
2. E. Hadicke, E.C. Mez, C.H. Krauch, G. Wegner, and J. Kaiser, Agnew. Chem., 83, 253 (1971).

3. D. Kobelt and H. Paulus, Acta Cryst., B30, 232 (1974).
4. A.J. Melveger and R.H. Baughman, J. Polymer Sci., A-2, 11, 603 (1973).
5. V. Enkelmann and J.B. Lando, manuscript in preparation.
6. D. Day and J.B. Lando, submitted, J. Polymer Sci.
7. K.B. Blodgett and I. Langmuir, Phys. Rev., 51, 964 (1937).
8. A. Cemel, T. Fort, Jr., and J.B. Lando, J. Polymer Sci, A-1, 10, 2061 (1972).
9. A. Cemel, Ph.D. Thesis, "The Preparation and Characterization of Ultrathin Polymeric Films and Membranes," Case Western Reserve University, 1974.
10. M. Puterman, T. Fort, Jr., and J.B. Lando, J. Colloid Interface Sci., 47, 705 (1974).
11. D. Naegele, H. Ringsdorf and J.B. Lando, manuscript in preparation.
12. V. Enkelmann and J.B. Lando, J. Polymer Sci., Al, in press.
13. N. Morosoff, H. Morawetz, and B. Post, J. Amer. Chem. Soc., 87, 3035 (1965).
14. G. Kumar, Ph.D. Thesis, Part II, The Crystal Structure of a Metastable Phase of N-p-Chlorophenyl Phthalanilic Acid and the Mechanism of Phase Change to the Stable Phase", Case Western Reserve University, 1974.
15. J.P. Mornon, Acta Cryst., B26, 1985 (1970).
16. B.L. Farmer and J.B. Lando, Z. Naturforsch., 26b, 769 (1974).
17. J.B. Lando and J. Semen, J. Polymer Sci., A-1, 10, 3003 (1972).

DISCUSSION

A. POCIUS - 3M CO. Have you conducted any studies comparing the rate of polymerization with the Langmuir-Blodgett films versus the rate in other systems of the same monomer? J. B. LANDO - CASE - WESTERN RESERVE UNIVERSITY. The rate of polymerization of the vinyl stearate multilayers is virtually identical with the rate of polymerization of bulk solid state polymerized material under exactly the same conditions of polymerization because it's the same crystal structure. The rate of polymerization of the head to head structure underneath the water surface is greater. We have looked at this system in relative detail.

A. POCIUS. Have you compared the molecular weight distribution from polymerization in Langmuir-Blodgett films to the distribution in other systems of the same monomer? J. B. LANDO. I am not sure about that. I know the molecular weight in the solid state polymerization, and also the molecular weight in multilayer polymerization, which we measured by GPC. They are very much higher than the molecular weights that we get by solution polymerization. I think it was about 150,000 to 160,000 based on a polystyrene standard so it is not really an accurate molecular weight but it is in the right ball park. It certainly gives you much higher molecular weight than you get in solution.

W. BAILEY - U. OF MARYLAND. I was curious about the volume changes that occur in these reactions. J. B. LANDO. The volume changes in the diacetylenes, especially when there is solid solution formation, is on the order of a percent or two, and, in general, not as high as 10%.

W. BAILEY. In the ones that had solid solutions, was there a volume change? J. B. LANDO. Yes there is, one to a few percent (and up to 10% - Ray Baughman). In the multilayers, with the paraffin packing, there is again very little change (we made calculations of this) in the packing density of the side chains, except when you go from the solid solution of polymer in monomer to the solid solution of monomer in polymer. There is very little change during the initial stages, e.g., the spacing, and this is a

monoclinic stacking in the side chain direction; d001, e.g.,
for vinyl stearate changes from 25.4 Å to 25.9 Å from 0 to
just before the phase change occurs at around 50% polymerization. There are larger changes, obviously, in a
nucleated reaction; larger changes are possible in those
cases and, of course, in the case of where you get amorphous
product. In the latter case, of course, you can have any
volume change. You have no direct crystallographic
correlation between a reactant and a product.

THE SOLID STATE SYNTHESIS AND PROPERTIES OF PHOTOCONDUCTING, METALLIC, AND SUPERCONDUCTING POLYMER CRYSTALS

Ray H. Baughman

Materials Research Center
Allied Chemical Corporation
Morristown, New Jersey 07960

INTRODUCTION

The electrical and optical properties of fully conjugated linear polymers have been of considerable research interest during the past several decades(1-3). The motivation has been to utilize the enormous variety of possible backbone and sidegroup structures to obtain new semiconductive, photoconductive, metallic, and superconductive materials.

The problem which has hindered both the development of materials applications and the understanding of structure-property relationships has been the ill-defined and multicomponent nature of conjugated polymers prepared using conventional techniques. The polymeric reaction products obtained, for example, by solution polymerization are either low molecular weight oligomers or low crystallinity, insoluble, and infusible polymers. Intractability, typical for polymers with conjugated backbones, severely limits methods of polymer processing, purification, and characterization. As a consequence, electrical measurements generally have had to be performed on compacted powders of polymers with poorly defined structures and high levels of impurities, such as reaction catalyst and reaction side products. Typically, inherent chain anisotropies in these polymers are averaged by structural disorder which occurs from the submicron down to the molecular scale. For these reasons it is not surprising that the electrical conductivities reported by different investigators for the same polymer sometimes differ by more than a factor of 10^5(2).

The discovery that certain fully conjugated polymers can be synthesized as large-dimension crystals has eliminated many of these barriers for fruitful research work. The synthetic method utilizes the three dimensional periodicity of a crystalline monomer phase as a template to determine the molecular and crystallographic structures of a polymer(4-12). In the ideal case, solid state reaction (initiated either thermally, mechanically, or by exposure to actinic radiation) transforms a monomer single crystal to a polymer single crystal with nearly the same dimensions and similar structural perfection. G. Wegner first demonstrated that certain diacetylene reactions closely approximate this ideal case(4).

The successful solid-state synthesis of large dimension single crystal polymers requires that a number of conditions should be satisfied. For this reason most solid-state reactions result in polymerization products which have much lower structural perfection than the precursor monomer crystals.

The first part of this work is concerned with the criteria which must be satisfied in order to synthesize polymer single crystals with high structural perfection by solid-state reaction. These criteria provide guidelines for selecting monomer phases likely to polymerize in the solid state by a reaction which preserves perfection and for understanding why particular reactions result in polymer phases containing specific types of structural disorder. The second part of this work deals with the host of unusual properties observed for conjugated polymers synthesized by solid-state reaction.

Two different kinds of solid-state reactions will be considered in some detail. These reactions are the 1,4-addition polymerization of acetylenic monomers containing the diacetylene functionality and the ring-opening reaction of the cyclic molecule S_2N_2 to produce the linear polymer $(SN)_x$. Because of backbone conjugation, the polymers produced by both types of reactions typically look metallic in appearance. However, while $(SN)_x$ is metallic at room temperature(13) and superconducting at low temperatures(14), the polydiacetylenes are large bandgap semiconductors and photoconductors(15-22).

CRITERIA FOR THE SYNTHESIS OF POLYMER SINGLE CRYSTALS

Least Motion Criteria

The observed polymerization mode for a particular monomer phase will generally be that mode which requires minimum atom displacements. As the atom displacements required for polymerization by the most favorable polymerization mode increase in going from one monomer phase to another, the phase polymerizability will tend to decrease. Consequently, if the atom displacements required for polymerization by all possible reaction modes are too large, no substantial polymerizability will be observed(7,11). Reaction, if it occurs at all, will be localized on structural defects such as dislocations, stacking faults, and crystal interfaces.

The basic approximation in the application of these least motion concepts to calculate solid-state reaction modes and relative phase reactivities is that the lattice contribution to the activation energy for reaction increases monotonically with the root mean square atom displacements required for reaction. For this reason only polymerization modes involving similar covalent bond changes can be compared. In using the least motion principle to calculate the structure of partially polymerized phases, an analogous approximation is made. This approximation is that the energy of the polymerizing phase will be lowest when the RMS atom displacements of the reacting molecules during polymerization are minimized.

For diacetylene phases and S_2N_2 the least motion concepts have been used to correctly predict reaction mode, relative phase reactivities, and the structure of partially polymerized phases(7,11,23,24). These concepts are useful for determining whether the likely packing modes for a particular monomer molecule result in a near neighbor configuration favorable for a specific type of reaction. For this purpose parameters are chosen to describe the important aspects of molecular packing in a possible reaction direction. Typical parameters are, for example, the monomer-monomer center-to-center separation and an angle which describes the orientation of possibly reacting functionalities. The RMS displacements (δ) required for polymerization are calculated as a function of these parameters. By examining the structures of analogous compounds with known reactivities, the maximum value of δ consistent with usable solid state reactivities (δ_m) can be established. Hence, the range of packing parameters consistent with useful phase reactivities (δ less than δ_m) is determined. Finally, packing constraints such as non-overlapping van der Waals volumes and possible hydrogen bonding requirements are examined in order to predict whether packing parameters consistent with useful phase reactivities are likely for specific compounds.

This approach is illustrated in Fig. 1 for the solid state polymerization of an array of diacetylene monomers to produce a polydiacetylene chain. The parameter δ is calculated for backbone associated carbon atoms as a function of the packing parameters d and γ, which are defined in Fig. 1.

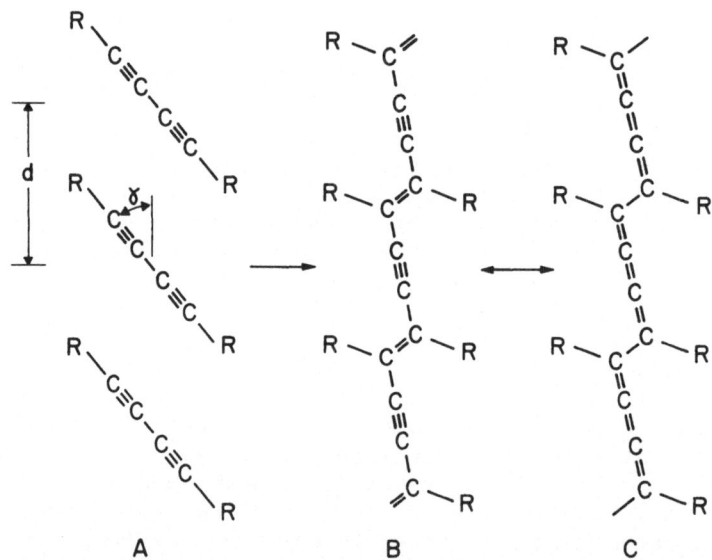

FIG. 1. Solid-state polymerization of an array of diacetylene molecules (A) to produce the 1,4-addition polymer described by the alternate mesomeric structures given in B and C. The molecular packing parameters relevant for evaluating phase reactivity (d and γ) are defined in A.

It is found that δ is less than 1.1Å for phases with suitable reactivity for polymer crystal synthesis(7,11). This limiting value of δ_m requires that γ be between 30 and 60° for general values of d and in a more limited angular range for specific values of d.

The utility of the least motion concepts is further illustrated by work on the solid-state synthesis of $(SN)_x$ from S_2N_2(23,24). Both the S_2N_2 phase and the final solid-state reaction product are monoclinic, with the same space group $(P2_1/c)$ and four SN units per unit cell(25-27). Consequently, since the polymer chains in $(SN)_x$ are in the unique axis direction (b), it was initially believed that reaction occurred in the crystallographic direction in S_2N_2. The least motion calculations indicate that formation of an all trans chain requires much larger δ than is required for formation of a cis-trans chain, which is in agreement with the observed cis-trans structure of $(SN)_x$. More important, these calculations indicate that a much smaller δ (0.48Å) is required for reaction in the a-axis direction than for reaction in the b-axis direction (1.20Å) or for any other possible reaction mode(23).

The calculated atom displacements for the predicted reaction mode are indicated in Fig. 2, which shows the a-axis array of mutually reacting S_2N_2 molecules and the resulting polymer chain. The chain atom positions calculated for partially polymerized S_2N_2 agree with the observed atom coordinates in this material, as determined by x-ray diffraction(27), to within a RMS deviation of 0.14Å(23). This correct prediction of reaction mode, chain backbone structure, and chain coordinates in the partially polymerized phase is a major success of the least motion calculations.

Uniqueness Criteria

The solid-state polymerization reaction of S_2N_2 is non-unique, which explains why $(SN)_x$ has much lower structural perfection than is obtainable for certain polydiacetylenes. In general terms, a solid-state transformation will be nonunique unless all symmetry elements of the monomer lattice are preserved in the product lattice. This means that a solid-state polymerization reaction will be non-unique unless (1) the monomer site symmetry is a possible

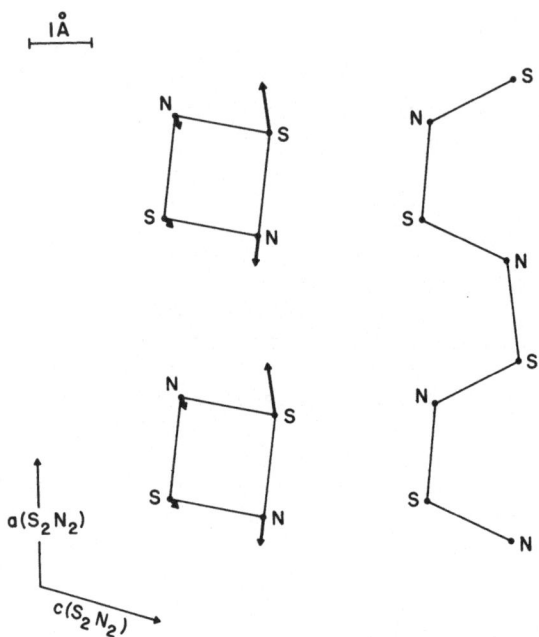

FIG. 2. Reaction mode predicted using the least motion
principle for the solid-state polymerization of S_2N_2 to
$(SN)_x$. The arrows indicate the atom displacements required
to transform the dimer array (left) into the polymer chain
(right). An equivalent reaction mode is obtained by in-
verting these displacement vectors through the molecular
center of symmetry.

site symmetry for the monomer unit in the polymer chain and
(2) the symmetry element relating mutually reacting
monomer molecules is a possible symmetry element between
the corresponding monomer units in the polymer chain(7).

    Because the first of these symmetry criteria is vio-
lated, the known S_2N_2 phase does not constitute an adequate
template for the solid-state synthesis of $(SN)_x$. Each S_2N_2
molecule sits on a lattice center of symmetry, which is not

a possible symmetry element for the S_2N_2 unit in the polymer chain. Because of this molecular center of symmetry,
opposite bonds in the S_2N_2 molecule are equivalent and,
therefore, are equally likely positions for ring cleavage
to form the polymer chain. Consequently, the reaction is
non-unique and proceeds by reaction at either of these bonds
to form distinguishable chains from different equivalent
molecular arrays throughout the crystal(23,24,27).

Depending upon the cooperativeness of solid-state reaction, such reaction non-uniqueness results in molecular
scale disorder, twinning, and a statistical relationship
between parent and product crystal orientations. For the
case of $(SN)_x$ synthesis, all of these effects are observ-
able(23-29).

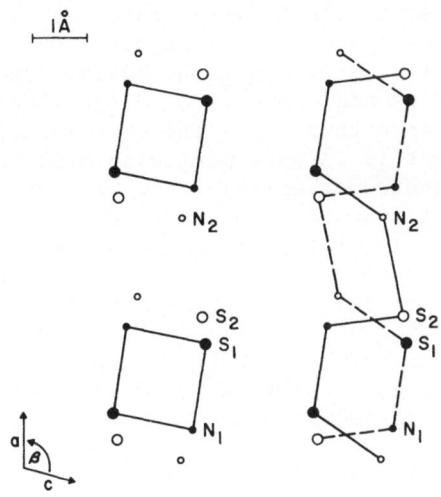

FIG. 3. Sulfur and nitrogen positions in partially polymerized disulfur dinitride (denoted S_1 and N_1 for 90% occupied sites and S_2 and N_2 for 10% occupied sites) in relationship to the superimposed contributions from the unpolymerized dimer arrays (left) and the centrosymmetrically
related $(SN)_x$ chains which result from reaction non-uniqueness (right).

The reaction non-uniqueness is first evident for the partially polymerized phase. As illustrated in Fig. 3, the electron density peaks determined by x-ray diffraction are a consequence of the contributions from unpolymerized S_2N_2 molecules and from the two different types of chains which are equally likely to form from each S_2N_2 array. As is expected from the small atom displacements calculated for one-half of the sulfur and nitrogen atoms (Fig. 2), these chain atom positions are unresolved from those of unreacted S_2N_2 molecules (Fig. 3).

As a consequence of the reaction non-uniqueness which results in formation of either of two chain types, the polymerizing S_2N_2 phase is progressively disordered. The statistical alternation of these chain types is reflected in disordered positions which appear in the electron density maps for the final polymer crystals(23,24).

A second example which indicates the importance of monomer phase symmetry for determining reaction uniqueness in polymer crystal synthesis is illustrated in Fig. 4. The known phase of the symmetric triyne with substituent groups $-CH_2OH$ is predicted using the least motion concepts to preferentially react by addition across diacetylene groups rather than across either acetylene or triacetylene groups (11). This aspect is in agreement with prior experimental work of Wegner and coworkers(8). However, the solid-state reaction of the triyne diol does not result in a single crystal polymer(11). The problem is that the point group symmetry of the monomer molecule in the observed phase is $\bar{1}$ (center of symmetry). Since this symmetry element is not possible for the monomer unit in the ordered polymer chain the reaction is non-unique. Solid-state reaction occurs non-periodically, as illustrated schematically in Fig. 4, to produce a disordered polymer phase.

In addition to reaction non-uniqueness, the occurrence of non-unique phase transformations during reaction is a possible origin of gross structural defects in polymerized phases. A solid-state phase transformation will be non-unique unless the symmetry elements of the parent lattice are included in the symmetry elements of the product lattice. If this condition is not satisfied, the product phase will be statistically disordered so as to produce an overall symmetry which is a sum of the symmetries of parent and product phases.

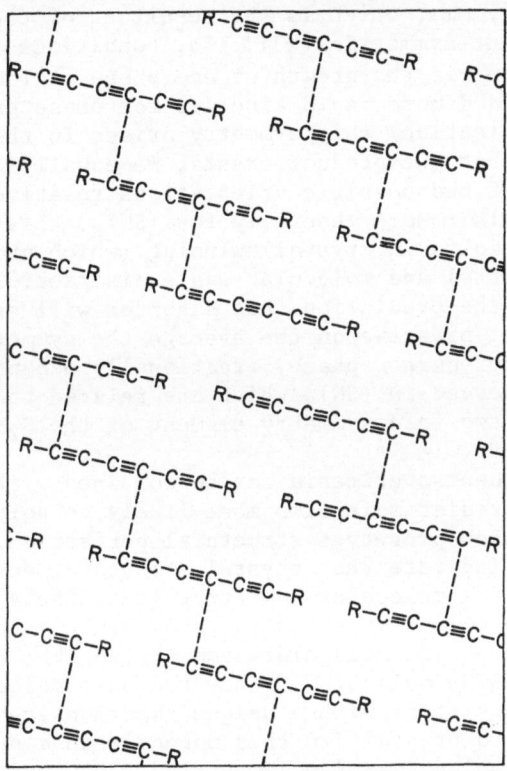

FIG. 4. Non-unique solid-state polymerization of a triyne
with molecular site symmetry $\bar{1}$. The dashed lines indicate
the mutually reacting carbon atoms.

The S_2N_2 reaction also provides a good example of the
occurrence and consequences of this type of non-uniqueness.
Although both dimer and final polymer phases have the same
space group ($P2_1/c$) and the same number of SN groups per
unit cell, the $2/m$ point symmetry elements of the two
phases are rotated with respect to each other by 90°. As a
consequence, the reaction product (SN)$_x$ has the overall

point symmetry mmm, which is the resultant of the two orthogonal 2/m point symmetries(23). For conditions favoring
transformation via the growth of one phase nucleus (small
crystal size and more rapid kinetics for phase growth than
for phase nucleation) this symmetry arises in the statistical sense that the product crystal is equally likely to
have either of two possible orientations relative to the
parent crystals. More generally for $(SN)_x$, this non-
uniqueness results in crystal twinning (which produces mmm
average symmetry) and molecular scale disorder(23). Consistent with the prediction that disorder will be introduced so as to preserve, on the average, the symmetry elements of the precursor phase, fractionally occupied defect
sites are observed in $(SN)_x$ which are related to the ordered
sites by the two-fold symmetry element of the S_2N_2 phase.

The uniqueness criteria can be combined with packing
concepts to predict molecules most likely to polymerize by
a reaction which preserves structural perfection. These packing concepts indicate that crystal structures are highly likely to preserve $\bar{1}$ molecular symmetry, less likely to preserve
either 2 or m molecular symmetry, and unlikely to preserve
mm, 2/m, 222, or mmm molecular symmetry(30,31). For example,
consider the S_2N_2 molecule. Since the only molecular point
symmetries consistent with a unique reaction are 1 and m, it
is desirable to crystallize this molecule in a phase having
one of these molecular site symmetries. However, molecules
like S_2N_2 having $\bar{1}$ symmetry are most likely to retain this
symmetry element in the crystalline state. On the other
hand, if one of the sulfur atoms in S_2N_2 were replaced by
selenium, the probability of discovering a phase consistent
with a unique solid state reaction would be greatly enhanced. Similarly, note that a unique reaction of a triyne
$R_1(C{\equiv}C)_3R_2$ by addition across a diacetylene group again requires a molecular site symmetry of 1 or m. Consequently,
the probability of finding a monomer phase suitable for
polymer crystal synthesis is increased by choosing $R_1{\neq}R_2$
(or R_1 chiral and identical to R_2) and either R_1 or R_2 incompatible with any symmetry other than 1 or m.

Note that the symmetry restrictions do not preclude the
synthesis of monocrystalline polymers having complex structural periodicities. For example, the monomer array symmetries (obtainable for orthorhombic or lower symmetry

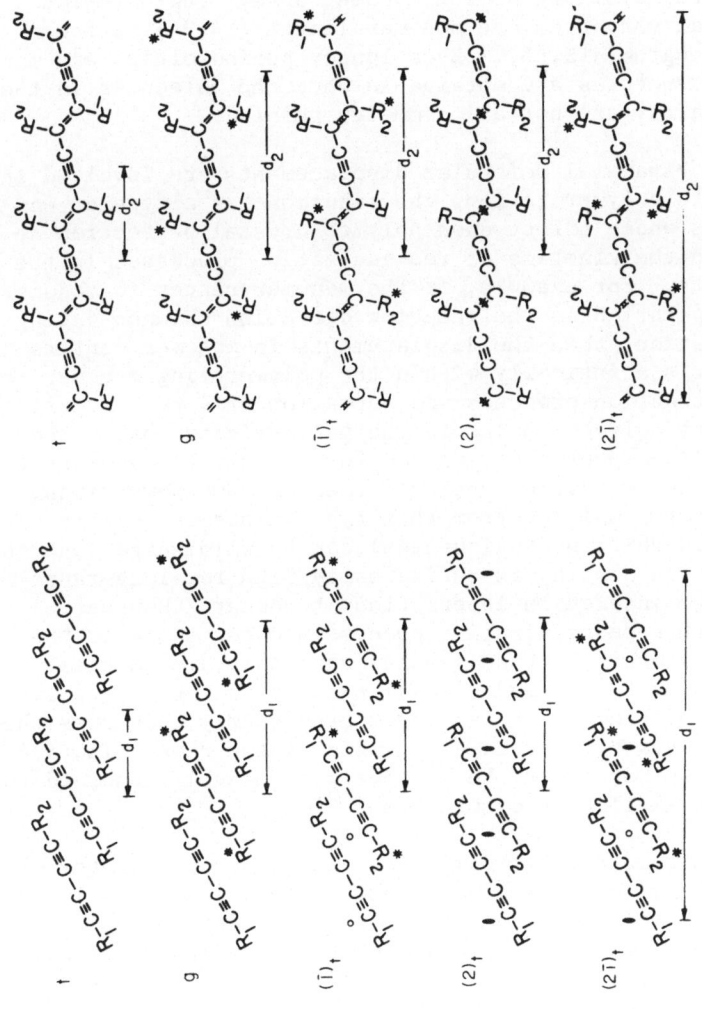

FIG. 5. Monomer array symmetries (left) consistent with unique solid-state reactions to form complex polymer chain structures (right). The asterisk denotes a change in handed-ness if the substituents are chiral.

phases) consistent with unique solid-state reactions are
pictured in Fig. 5. The symbols on the left denote the
sequence of symmetry elements which generates the monomer
array (t,g,2, and $\bar{1}$, respectively, for lattice translation,
glide plane, two-fold axis, or center of symmetry). Depending upon the symmetry of the monomer array, chain repeat
lengths can vary from one monomer length (~4.9Å) to four
monomer lengths (~19.6Å). Even longer periodicities and more
complex structures are obtainable when the molecules in the
reacting array are not all symmetry related.

If substantial molecular displacements are involved in
solid-state polymerization, the reaction has a type of non-
uniqueness whose effect upon polymer crystal perfection depends upon the kinetics of reorganization processes in the
solid state. For example, if the monomer-center to monomer-
center separation in the reaction direction changes during
polymerization, then the displacements in monomer centers
can differ statistically within the polymerizing monomer array. If reaction proceeds via formation of a random solid
solution of polymer chains in the polymerizing phase, these
shifts will be poorly correlated for essentially non-inter-
acting (well separated) chains formed at low conversions.
The disorder resulting from this type of non-uniqueness
(designated shift non-uniqueness) can be eliminated when the
polymer chain density is sufficiently high for long range
cooperative interchain interactions to occur. However,
whether these reorganization processes occur so as to produce a well ordered phase depends upon molecular mobility
at these conversions and the existence of a periodic polymer chain packing which has lower free energy than does the
disordered state. For cases in which these reorganization
processes do not occur the polymerized phase is observed to
be characteristically disordered.

The structures typically obtained as a consequence of
shift non-uniqueness are analogous to nematic phase liquid
crystals in the sense that there exists a unique chain direction, but the monomer repeat units in neighboring chains
are not well correlated in longitudinal position. As a
consequence, the x-ray rotation photographs about the chain
axis direction (c) display either (1) sharp hk0 reflections
and progressively more diffuse hkℓ reflection with increasing ℓ (2) diffuse reflection for all layers combined with a
few sharp hk0 reflections or (3) diffuse reflections on

defined layer lines. These situations result from non-unique reactions with unfavorable reorganization kinetics in which, respectively, monomer crystal periodicity is preserved (1) in the chain axis projection (2) in one direction within the chain axis projection and (3) only in the sense that there exists a common chain axis direction and monomer repeat length.

Phase Stability Aspects

In order to synthesize polymer single crystals with high structural perfection it is important that the polymer enters the polymerizing lattice as a solid solution over the entire monomer-to-polymer conversion range. This is because phase separation during reaction can produce enormous shear and volume strains which result in fragmentation or local deformation of the polymerizing phase(5,7). Furthermore, as discussed in the last section, unless the daughter crystal has high enough symmetry to include the symmetry elements of the parent crystal these symmetry elements will be retained via the introduction of disorder(32).

The x-ray diffraction results indicate that the $(SN)_x$ chains enter the polymerizing S_2N_2 phase as a solid solution over a significant conversion range. However, at a higher conversion (which probably depends upon crystal size, crystal inhomogeneities, and reaction temperature) phase separation occurs. This phase separation occurs as the only process whereby a crystalline polymer can result from the observed polymerization mode of the known dimer phase.

Phase separation occurs during certain diacetylene reactions, despite the fact that a continuous monomer-to-polymer single crystal transformation is not forbidden by symmetry considerations. For example, the diacetylene with substituent groups $-CH_2OH$ can be polymerized to a limiting conversion of about 70% as a one-phase reaction. Annealing this partially polymerized phase results in phase separation to produce a non crystalline polymer phase and the initial monomer phase(6).

The occurrence of limited conversion, phase separation, and shift non-uniqueness are often related. The greater the difference between the structure of the polymer chain and that of the precursor monomer array, the higher the lattice

strain energy which will favor limited conversion and phase
separation and the greater the likelihood of important shift
non-uniqueness. The former aspects can be quantitatively
considered using the theory of binary solid solutions to
construct free energy diagrams for polymerizing phases(7).

These considerations have important consequences with
respect to the choice of monomer. For polymerization reactions involving large van der Waals volume change of the
reacting groups, small or inflexible substituent groups are
unfavorable for polymer crystal synthesis. According to
the group increments tabulated by Bondi(33), the diacetylene polymerization requires about an 18% change in the
total van der Waals volume of the backbone associated carbon
atoms. In order for this van der Waals volume change on
polymerization to be accommodated by the polymerizing phase
without resulting in large dimensional changes and high
lattice strains, it is most reasonable to choose a monomer
phase in which there are long flexible sidegroups and/or interstitial molecules. Unfortunately, this choice which is
convenient for successful polymer crystal synthesis also reduces the polymer chain density. Hence, chain-associated
properties are correspondingly reduced. A similar problem
does not arise in the S_2N_2 ring-opening polymerization because there is no increase in the number of covalent bonds.

The liklihood of transformation non-uniqueness is increased for monomers containing more than one potentially
reactive group. For example, the cyclic diacetylene shown
in Fig. 6 is obtained as a reactive phase containing interstitial chloroform by crystallization from a petroleum
ether/chloroform solution(34). Raman and x-ray diffraction
results establish that polymerization occurs by 1,4-addition
of diacetylene groups to produce a chain-aligned polymer.
However, as previously described for other non-unique diacetylene reactions, the polymer is crystallographic only
in the chain-axis projection. For the case that reaction
occurs at each diacetylene functionality, as indicated in
Fig. 6, the multiple connectivity of the resulting ladder
polymer and the large van der Waals volume change per monomer unit favors such a non-unique reaction. Because of the
relative insensitivity of resonance Raman, infrared, and
ultraviolet spectra to the presence of unreacted diacetylene groups, it has not yet been established whether chains
are formed from each of the four diacetylene groups in the
monomer molecule. For reactions such as this, which result

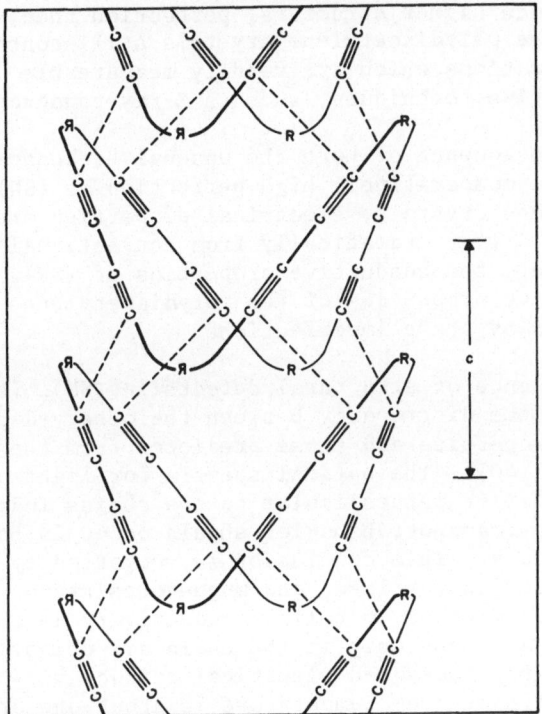

FIG. 6. Schematic representation of the structure and possible reaction mode of a cyclic tetradiyne. Reacting carbon atoms are connected by dashed lines. R is $(CH_2)_4$.

in an insoluble and infusible polymer which does not have three-dimensional periodicity, it is difficult to fully characterize associated structural changes.

POLYMER CRYSTAL PROPERTIES AND POLYMER PERFECTION

It is important not to lose perspective when considering the defect sensitive properties of polydiacetylene and $(SN)_x$ crystals. Although $(SN)_x$ produced by solid-state reaction is not well ordered. the crystalline perfection of this polymer is much higher than for conventional organic polymers. Polydiacetylene crystals are obtainable

which have much higher structural perfection than $(SN)_x$.
However, these polydiacetylene crystals still contain defect concentrations which are readily measureable using
highly sensitive techniques, such as X-ray topography(35-36).

As a consequence of both the unusual backbone structures and the comparatively high perfection of $(SN)_x$ and
polydiacetylene crystals, electrical properties are observed which differ dramatically from conventional polymers. However, the conductive properties of $(SN)_x$ and the
photoconductive properties of the polydiacetylenes are
still limited by phase imperfections.

The presence of structural defects in $(SN)_x$ is believed
to result in the discrepancy between the observed dc
electrical properties and those predicted from the optical
spectra(13,37-40). The optical spectra for light polarized
either parallel or perpendicular to the chains indicate
that metallic transport behavior should occur in both of
these directions. This conclusion is supported by thermo-
electric power, energy loss, and magnetoresistance measure-
ments(41-43). However, metallic conductivity is observed
by direct measurement only in the chain direction. Furthermore, the highest measured electrical conductivity in the
chain direction at room temperature is about one order of
magnitude lower than that calculated by Drude analysis of
the optical data(38,44). These discrepancies, as well as
the general shape of the absorption spectra, can be explained by assuming that $(SN)_x$ consists of metallic fibers
which are partially insulated with respect to one another(44).

With decreasing temperature the electrical conductivity
of $(SN)_x$ in the chain direction increases and the orthogonal
conductivity decreases slightly(13,37-40). Greene, Street,
and Suter discovered that at still lower temperatures a
transition occurs which results in superconductivity in all
crystal directions(14). As is true for polymeric sulfur
nitride in the metallic state, the superconductive properties are dependent upon structural imperfections in $(SN)_x$
(38-40,45-47). The transition temperature, T_c, increases and
the transition sharpens for crystals with higher resistivity
ratios, $\rho(300^\circ K)/\rho(4^\circ K)$, and, presumably, higher structural
perfection(39,47). For the poorer quality crystals, low
critical field values and substantial remnant resistivities
below the transition are observed(45-47). Similarly, the

high observed anisotropy of critical field appears to result from irregularities in the lateral packing of chains,
which define an effective fiber diameter(39,45). Experiments relevant to superconducting fluctuations suggest
that the T_c for a defect-free crystal of $(SN)_x$ should be
near 1.4°K, or about four times higher than the highest
value presently observed (0.35°K)(40,48).

The observed defect dependence of electrical properties and the crystal-to-crystal variation in these properties for $(SN)_x$ is consistent with the previous arguments
that this polymer should contain gross structural defects.
This property variation is likely to be a consequence of
differing degrees of cooperative reorganization during the
dimer-to-polymer transformation. The extent to which the
reaction and phase transformation non-uniqueness will be
reflected in a statistically determined polymer crystal
orientation, rather than via molecular scale disorder or
twinning, will depend upon the cooperativeness of the transformation and the occurence of related reorganization processes. More specifically, the cooperativeness of polymerization will depend upon the chain length resulting at a
particular conversion from a single chain initiation event.
Similarly, the cooperativeness of the phase transformation
will depend upon the relative kinetics of crystal nucleation
and crystal growth. All of these factors are expected to
depend upon polymerization temperature and dimer crystal
perfection. Dimer crystal impurities and imperfections
could serve as chain initiation and termination sites or as
crystal nucleation sites. The conversion at which phase
nucleation and growth occurs is important because further
polymerization will be strongly effected.

Relevant to these ideas, radicals are observed in
polymerizing S_2N_2 and not in the final polymer(27). These
radicals are believed to correspond to chain ends. This
is consistent with the presence of chain initiation events
which do not result in polymerization across the entire
crystal dimension. Secondary chain initiation events in
the same monomer array are, in the absence of interactions,
equally likely to produce either of the two chains shown in
Fig. 3. At higher conversions, when these independently
produced chain segments grow together or are linked via
the initiation and growth of interconnecting chain segments,
point defects must be produced at the junctions between

chain segments having centrosymmetrically related structures. The resulting point defect density, on a volume or
weight basis, is one-third of the density of independent
chain initiation sites. Note that neighboring chain ends
corresponding to array segments which have polymerized by
bond opening at centrosymmetrically related SN bonds must
be either both sulfur or both nitrogen. This means that
mutual reaction of these chain ends will produce either S-S
or N-N chain linkages as chemical defects in the chain. Al-
ternately, the chain end radicals could terminate via reaction with chemical impurities present in the crystals.
Such termination would be consistent with the high observed
hydrogen impurity levels in $(SN)_x$(49). In either case, this
aspect of reaction non-uniqueness is expected ot result in
defect-associated electronic states and scattering centers
in $(SN)_x$.

In contrast to $(SN)_x$, both optical data and dc conductivity measurements indicate that the polydiacetylenes
are large bandgap semiconductors(15,16,19,50-54). Investigation of model compounds suggest that the band gap in the
polydiacetylenes can be approximated by $E_g=E_\infty+C\ell^{-1}$, where
E_∞ and C are constants and ℓ is the backbone conjugation
length(53,54). Chain ends, chemical impurities, and configurational defects can limit ℓ. However, even in the absence of non-equilibrium defects, chain conjugation will be
interrupted by a finite concentration of entropically
stabilized point defects(54-56). The smallest observed
bandgap for a single crystal polydiacetylene(50-52) and the
predicted bandgap for an infinite chain(53,54) are both about
2.0eV. This suggests that chain defects are not so numer-
ous in the best of these crystals to substantially change
the band gap.

The dark field conductivities of most polydiacetylenes
are so low that it is uncertain to what extent the measured
values correspond to a crystal property(15-17). The lowest
measured room temperature values are about 10^{-15}ohm^{-1}cm^{-1}
(17), or about 10^{-18} times the chain direction conductivity of $(SN)_x$. We see that the similar metallic appearance of these two different types of polymers is
certainly not indicative of analogous electronic properties.

The photoconductive properties of the polydiacetylenes
are much more interesting than the dark field conductive

properties(15,17-22). Transient photoconductivity measure-
ments(17,22) indicate that the carrier mobility of both
holes and electrons for a polydiacetylene (substituent
groups $-CH_2OSO_2C_6H_4CH_3$) is about 1 cm^2/volt sec., or about
10^7 times higher than for the commercially important polymeric
photoconductor, poly (N-vinylcarbazole). Such a mobility
value is similar to those observed for many organic molecular
crystals. Severe trapping, corresponding to a carrier lifetime of ~1µs for both holes and electrons, is observed for
even the best polymer crystals. Polymerization via gamma
ray irradiation, as opposed to thermal annealing, substantially decreases these carrier lifetimes. Assuming
reasonable values for the carrier thermal velocity and a
typical neutral trap cross section, a 1µs carrier lifetime
corresponds to a deep trap density of ~10^{14} cm^{-3}, or about
one deep trap per 10^7 monomer units(17). We see that small
defect concentrations which would have no measurable effect
on optical properties (band gap) can result in enormous
changes in photoconductivity.

Polydiacetylene crystals are found to have unusually high third-order optical susceptibilities, χ_3, for
light polarized in the chain direction(57). In other
words, one-dimensional π electron delocalization in the
polymer backbone results in a chain direction polariz-
ability which is significantly dependent on the second
power of electric field. For the investigated centrosymmet-
ric polymers, higher order susceptibilities are too small to
be of interest and the second-order susceptibility is con-
strained by symmetry to be zero. Observed values of χ_3 for
polydiacetylenes are about four orders of magnitude higher
than for any other investigated material with a comparable
transparency range in the visible. This fact coupled with
the high observed radiation damage thresholds for pico-
second light pulses, suggest the utilization of these
materials for the construction of ultrafast light shutters
(58) and four wave parametric amplifiers. Sauteret and coworkers have calculated that for a 1.06µm, 1GW/cm^2 pump,
the gain for four-wave parametric amplification in the near
infrared using a polydiacetylene crystal is about ten times
higher than the gain for three-wave parametric amplification
using more conventional materials such as $LiNbO_3$(57). Such
devices are of interest for obtaining high efficiency, vari-
able-frequency laser sources.

The mechanical properties of the polydiacetylenes reflect the high structural perfection obtainable for these
crystals. Because of relatively weak van der Waals or hydrogen bonding normal to the chain direction, brittle failure occurs at low stress levels for stresses applied normal
to the chain direction. However, the per chain mechanical
properties in the chain direction are spectacularly high
(59). The investigated polydiacetylene crystals have large
substituent groups which decrease the chain density, while
contributing insignificantly to per chain mechanical properties. As discussed earlier, upon decreasing the size of
these substituent groups so as to increase the overall tensile properties, it becomes progressively more difficult to
maintain structural perfection in the polymerized phase. In
other words, as the chain density is increased by decreasing the concentration of non-backbone atoms, the polymer
crystals tend to become less perfect, corresponding to effectively weaker chains.

Mechanical properties have been evaluated for
single crystal fibers of the dioxane-containing phase of a
polydiacetylene with substituent groups $-CH_2OCONHC_6H_5$(59).
The observed properties are similar to those of metal and
ceramic whiskers. The elastic modulus is strain dependent
and the ultimate strength increases with decreasing crystal
size. The latter behavior suggests that the ultimate tensile strength of these fibers is controlled by the presence
of a small number of defects at which fracture nucleates.
Upon decreasing the fiber volume and surface area the ultimate strength increases because the probability that a
critical size flaw (required for fracture nucleation at a
particular stress level) exists within the sample decreases.

The per chain modulus of this polymer is about equal
to that of diamond in the [110] direction. A polyethylene
fiber with the same per chain mechanical properties would
have an ultimate tensile strength in excess of one million
psi. The theoretical modulus calculated for a defect free
polydiacetylene chain using a spectroscopic force field is
within 10% of the observed modulus. This contrasts with
the case for conventional polymers, where the bulk tensile
modulus is typically much less than 50% of the theoretical
(spectroscopic) modulus.

We see that diverse properties of polydiacetylene

and $(SN)_x$ crystals reflect the high degree of crystal perfection, as well as the presence of crystal imperfections. Work in progress at various laboratories is concerned with increasing the structural perfection obtainable for existing polymer crystals, as well as the synthesis of new types of single crystal polymers designed to optimize particular properties. The unique polymer properties which have been obtained for polymer single crystals justifies this substantial effort.

References

1. Ya.M. Paushkin, T.P. Vishnyakova, A.F. Lunin, and S.A. Nizova, "Organic Polymeric Semiconductors", John Wiley and Sons, New York, 1974.

2. F. Gutmann and L.E. Lyons, "Organic Semiconductors", John Wiley and Sons, New York, 1967.

3. H. Mark, Israel J. Chem. 10, 407 (1972).

4. G. Wegner, Z. Naturforsch. 24b, 824 (1969).

5. J. Kaiser, G. Wegner, and E.W. Fischer, Israel J. Chem. 10, 157 (1972).

6. R.H. Baughman, J. Appl. Phys. 43, 4362 (1972).

7. R.H. Baughman, J. Polym. Sci., Polym. Phys. Ed. 12, 1511 (1974).

8. J. Kiji, J. Kaiser, G. Wegner, and R.C. Schulz, Polymer 14, 433 (1973).

9. R.H. Baughman and K.C. Yee, J. Polym. Sci., Polym. Chem. Ed. 12, 2467 (1974).

10. G. Wegner, Chimia 28, 475 (1974).

11. R.H. Baughman and K.C. Yee, J. Macromol. Sci.-Rev. Macromol. Chem., in press.

12. G. Wegner, Makrom. Chem. 145, 85 (1971).

13. V.V. Walatka, M.M. Labes, and J.H. Perlstein, Phys. Rev. Lett. 31, 1139 (1973).

14. R.L. Greene, G.B. Street, and L.J. Suter, Phys. Rev. Lett. 34, 577 (1975).

15. W. Schermann and G. Wegner, Makromol. Chem. 175, 667 (1974).

16. D. Bloor, D.J. Ando, F.H. Preston, and G.C. Stevens, Chem. Phys. Lett. 24, 407 (1974).

17. R.R. Chance, R.H. Baughman, P.J. Reucroft, and K. Takahashi, Chem. Phys. 13, 181 (1976).

18. R.R. Chance and R.H. Baughman, J. Chem. Phys. 64, 3889 (1976).

19. K. Lochner, B. Reimer, and H. Bässler, Chem. Phys. Lett. 41, 388 (1976).

20. B. Reimer and H. Bässler, Phys. Stat. Sol. 32a, 435 (1975).

21. K. Lochner, B. Reiner, and H. Bässler, Phys. Stat. Sol., in press.

22. B. Reimer and H. Bässler, private communication.

23. R.H. Baughman, R.R. Chance, and M.J. Cohen, J. Chem. Phys. 64, 1869 (1976).

24. R.H. Baughman and R.R. Chance, J. Polym. Sci., Polym. Phys. Ed. 14, 2019 (1976).

25. M. Boudeulle, Ph.D. thesis, Univ. of Lyon, 1974.

26. M. Boudeulle, Cryst. Struct. Comm. 4, 9 (1975).

27. J. Cohen, A.F. Garito, A.J. Heeger, A.G. MacDiarmid, C.M. Mikulski, M.S. Saran, and J. Kleppinger, J.A.C.S. 98, 3844 (1976).

28. A.G. MacDiarmid, C.M. Mikulski, M.S. Saran, P.J. Russo, M.J. Cohen, A.A. Bright, A.F. Garito, and A.J. Heeger, Advances in Chemistry Series, No. 150 (R.B. King, Ed.), 1976, p. 63.

29. G.B. Street, H. Arnal, P.M. Grant, and R.L. Greene, Mat. Res. Bull. 10, 877 (1975).

30. A.I. Kitaigorodsky, "Molecular Crystals and Molecules", Academic Press, New York, 1973, Chapter 1, pp. 33-37.

31. B. Wunderlich, "Macromolecular Physics", Vol. I, Academic Press, New York, 1973, Chap. II, pp. 21-61.

32. J.Z. Gougoutas, Israel J. Chem. 10, 395 (1972).

33. A. Bondi, "Physical Properties of Molecular Crystals, Liquids, and Glasses", John Wiley and Sons, New York, Chap. 14, 450.

34. R.H. Baughman and K.C. Yee, U.S. Patent 3,923,622 (Dec. 2, 1975).

35. J.M. Schultz, J. Materials Sci. 11, 2258 (1976).

36. R.E. Green, private communication.

37. A.A. Bright, M.J. Cohen, A.F. Garito, A.J. Heeger, C.M. Mikulski, and A.G. MacDiarmid, Appl. Phys. Letters 26, 612 (1975).

38. H.P. Geserich and L. Pintschovius, Festkörperprobleme (Advances in Solid State Physics) XVI, 65, J. Treusch (ed.), Vieweg, Braunschwieg (1976).

39. R.L. Greene and G.B. Street, Proceedings of the NATO-ASI on "Chemistry and Physics of One-Dimensional Metals", Bolzano, Italy, August 1976 (edited by H.J. Keller, Plenum Press, 1977).

40. G.B. Street and R.L. Greene, IBM Journal of Research and Development, in press.

41. M.J. Cohen, C.K. Chiang, A.F. Garito, A.J. Heeger, A.G. MacDiarmid, and C.M. Mikulski, Bull. Am. Phys. Soc. 20, 360 (1975).

42. C.H. Chen, J. Silcox, A.F. Garito, A.J. Heeger, and A.G. MacDiarmid, Phys. Rev. Lett. 36, 525 (1976).

43. W.D. Gill, W. Beyer, and G.B. Street, to be published in Proceedings of Siofok Conf. 1976 and Solid State Comm.

44. H. Kahlert and K. Seeger, "Electrical Properties of Polysulfur Nitride, (SN)$_x$", Proc. 13th Int. Conf. Phys. Semicond., Rome, Aug. 1976.

45. L.J. Azevedo, W.G. Clark, G. Deutscher, R.L. Greene, G.B. Street, and L.J. Suter, Solid State Comm. 19, 197 (1976).

46. R. Civiak, C. Elbaum, W. Junker, C. Gough, H.I. Kao, L.F. Nichols, and M.M. Labes, Solid State Comm. 18, 1205 (1976).

47. R. Civiak, W. Junker, C. Elbaum, H.I. Kao, and M.M. Labes, Solid State Comm. 17, 1573 (1975).

48. R.L. Civiak, C. Elbaum, L.F. Nichols, H.I. Kao, and M.M. Labes, submitted to Phys. Rev.

49. R.D. Smith, J.R.W. Wyatt, J.J. DeCorpo, F.E. Saalfield, M.J. Moran, and A.G. MacDiarmid, Chem. Phys. Lett., 41, 362 (1976).

50. D. Bloor, F.H. Preston, and D.J. Ando, Chem. Phys. Lett. 38, 33 (1976).

51. B. Reimer, H. Baessler, J. Hesse, and G. Weiser, Phys. Stat. Sol. (b), in press.

52. C.J. Eckhardt, H. Müller, J. Tylicki, and R.R. Chance, J. Chem. Phys. 65. 4311 (1976).

53. G J. Exarhos, W.M. Risen, Jr., and R.H. Baughman, J. Am. Chem. Soc. 98, 481 (1976).

54. R.H. Baughman and R.R. Chance, J. Polym. Sci., Polym. Phys. Ed. 14, 2037 (1976).

55. G.C. Stevens and D. Bloor, J. Polym. Sci., Polym. Phys. Ed. 13, 2411 (1975).

56. R.H. Baughman, G.J. Exarhos, and W.M. Risen, Jr., J. Polym. Sci., Polym. Phys. Ed. 12, 2189 (1974).

57. C. Sauteret, J.P. Hermann, R. Frey, F. Pradere, J. Ducuing, R.H. Baughman, and R.R. Chance, Phys. Rev. Lett. 36, 956 (1976).

58. J.P. Hermann, D. Ricard, and J. Ducuing, Appl. Phys. Lett. 23, 178 (1973).

59. R.H. Baughman, H. Gleiter, and N. Sendfield, J.
Polym. Sci., Polym. Phys. Ed. 13, 1871 (1975).

DISCUSSION

J. K. STILLE - U. OF IOWA. Did I hear you correctly in the beginning of the talk when you said you can obtain polymer crystals as long as 20 cm? R. BAUGHMAN - ALLIED CHEMICAL CO. There is no limit to the size polymer crystals which can be obtained, since the polymer crystals have essentially the same dimensions as the precursor monomer crystals. The 20 cm crystal referred to was obtained by Czochralski type melt growth of the precursor monomer crystal. In this case, monomer crystal growth continued until essentially the entire melt was consumed.

J. K. STILLE. Is this direction the direction of propagation of the chain axis? Do you have molecular weights corresponding to the length of 20 cm? R. BAUGHMAN. In a macroscopic crystal of centimeter dimensions, you do not have chains which proceed uninterrupted from one end of the crystal to the other. Chemical impurities and dislocation cores can interrupt polymerization, thereby producing a final structure which has statistically distributed chain ends. Because of the insolubility typical of many polydiacetylenes, it is difficult to evaluate molecular weight. However, by placing flexible sidegroups on the polymer backbone the solubility required for molecular weight determination can be obtained. Gel permeation chromatography on these polymers indicate high molecular weights, as does solution viscosity. For well ordered polydiacetylene crystals the macroscopic modulus is equal to the theoretical modulus, again indicating that the molecular weight is high.

J. K. STILLE. What kind of thermodynamic data do you get out of the polymerization reaction, particularly what are the entropy changes? R. BAUGHMAN. Using Dewar's tabulated values for bond energies as a function of bond order and bond hybridization, we have calculated that the molecular energy change for 1,4-addition polymerization of diacetylenes should be about -30 kcal/mole. The total enthalpy change during reaction is this molecular energy plus a substantial lattice energy contribution which is likely to have an opposite sign. Since the latter energy depends upon the crystal structures in the monomer and polymer phases, the overall heat of reaction is expected to vary even for different crystallographic phases of the same monomer. The molecular entropy term will decrease in going from monomer to polymer. However, for cases in which the molecular weight at low conversions is low, a very important mixing entropy term will affect reaction kinetics at these conversions. The same mixing entropy term can prevent reaction of the last traces of monomer in the polymerized crystal.

F. KARASZ - U. OF MASSACHUSETTS. Talking about $(SN)_x$, I have a question similar to the one that was raised just now. As I understand it $(SN)_x$ is totally insoluble, it also sublimes, and you can recover essentially what you started out with. Now on that basis what is the evidence that $(SN)_x$ is a polymer rather than oligomer of say, 8 or 10 units? R. BAUGHMAN. Measurements by Saalfeld and coworkers indicate that $(SN)_x$ sublimes by reversible bond cleavage to form a non-cyclic S_4N_4 species. Consequently, the fact that $(SN)_x$ sublimes appears to have no direct relationship to the polymer molecular weight. Because of polymer insolubility and the absence of information about end group character, no direct measurements of polymer molecular weight are available. The long Ginzburg-Landau coherence length in the chain direction, calculated by Azevedo and coworkers from critical field measurements for the superconducting transition, are consistent with a high molecular weight polymer. The chain direction mechanical properties are also consistent with a high molecular weight polymer.

F. BAILEY - UNION CARBIDE CORP. Could you give the Young's modulus in some relevant units? R. BAUGHMAN. The Young's modulus for a polymer which has a cross-sectional area per chain of about 5.3X that of polyethylene and does not have any hydrogen bonding in the chain direction is about 6.8 million lbs/in^2. This polymer has bulky substituent groups which result in this large cross-sectional area per chain and, correspondingly, reduce the measured modulus. However, the per chain modulus is extraordinarily high. For example, a polyethylene fiber with the same per chain modulus as the polydiacetylene would have a Young's modulus of 5.3 x 6.8 million lbs/in^2. The obvious question is, why not decrease the size of the substituent groups so that one can utilize to a greater extent the high per chain properties. The problem is that as the substituent groups

become smaller and smaller it becomes more and more difficult
to maintain structural perfection during solid state reaction.

D. ULRICH - AIR FORCE OFFICE OF SCIENTIFIC RESEARCH.
Have you tried to grow the polydiacetylenes by melt zone
refining for directional solidification?   R. BAUGHMAN.
Large polydiacetylene crystals can be obtained from monomer
crystals grown using zone refining techniques only if a trick
is employed. The problem is that diacetylene crystals
suitable for the synthesis of polydiacetylenes are usually
thermally reactive in the solid state. Since the monomer
melts are typically much less reactive, the crystal growth
methods which we use involve rapid heating of the monomer
from room temperature to the melt. This is done to avoid
unwanted polymerization prior to crystal growth, which would
interfere with homogeneous crystallization. During the zone
pass, the diacetylene phase will be heated in the solid state
for a substantial time period prior to melting and
recrystallization. To avoid reaction prior to crystal
growth, one can employ a thermally unreactive phase which
crystallizes from the melt as a thermally reactive phase.
This is feasible for several different diacetylenes because
of the high degree of polymorphism typically evidenced by
these monomers. Since the monomer is polymerizable after
recrystallization, only one zone pass could be made.

R. NATARAJAN - LORD CORP. Do you have any information
on the transverse properties of the polydiacetylenes?
R. BAUGHMAN. With the exception of the polydiacetylene
discussed by Jerry Lando, most of the polydiacetylenes which
have been investigated are hydrogen bonded transverse to the
chain direction. The mechanical properties normal to the
chain direction are dependent upon the anisotropy introduced
by this hydrogen bonding. Many property aspects are an
order of magnitude or more lower normal to the chain than
in the chain direction. Such properties include, for example,
elastic modulus, ultimate strength, and carrier mobility.

C. OVERBERGER - U. OF MICHIGAN. In the data you showed
on whiskers (5 cm long), were they made from the cyclic
material with the preacetylenic linkages, or were they made
by the more simple procedures?   R. BAUGHMAN. The
mechanical property data are for single crystal fibers of
the polydiacetylene derived from the solid-state reaction
of the diacetylene with substituent groups - $CH_2OCONHC_6H_5$.

C. OVERBERGER. What is the tensile data on that ladder structure?   R. BAUGHMAN. The modulus of the polymer derived by solid-state reaction of the cyclic tetradiyne is much lower than would be expected if the theoretical per chain modulus was achieved for each chain involved in a ladder structure. Since this polymer is crystallographic only in the chain axis projection, the low modulus is likely to be a consequence of the disorder. Limited chain lengths or unreacted groups also might contribute to the low modulus.

J. R. SCHAEFGEN - DU PONT. Do you have any X-ray, sonic modulus or some other measure of orientation in the chain direction? It should be almost perfect. R. BAUGHMAN. For a well ordered polydiacetylene single crystal there is a unique chain axis direction. No measurable deviation of individual chains from this direction is observed. The extraordinarily high per chain modulus and ultimate strength, the negative macroscopic thermal expansion typical in the fiber direction, and the optical anisotropy are all consistent with this conclusion. However, the most definitive evidence is provided by X-ray single crystal measurements. The reflections are extraordinarily sharp, as compared to those for conventional polymers, and are observable at very high Bragg angles. Jerry Lando mentioned that he was able to record over 800 reflections for one of the polydiacetylenes which he investigated.

R. F. KIESEL - FORD MOTOR CO. Do you have any idea of the hole size in that cyclic tetradiyne? Can you predict what kind of electronic structure and molecular size material would fit in it?   R. BAUGHMAN. We don't know precisely the structure of this polymer, so we can only estimate the size molecule which can be incorporated into the interstitial spaces. For example, chloroform would fit into the proposed structure, albeit with substantial strain energy.

R. F. KIESEL. Could you vary the hole size by varying the diacetylene?   R. BAUGHMAN. The size of the interstitial spaces could be increased by increasing the length

of the ring segments separating the diacetylene functionalities, as by increasing the number of methylenes. Since
interstitial molecules would be in close proximity to the
conjugated polymer backbones, this presents the possibility
of forming complexes in which there are interesting charge
transfer interactions.

ACETYLENE CONTAINING AROMATIC HETEROCYCLIC POLYMERS

F.E. Arnold and F.L. Hedberg

Nonmetallic Materials Division, Air Force
Materials Laboratory, Wright-Patterson Air
Force Base, Ohio 45433

R.F. Kovar*

University of Dayton Research Institute
Dayton, Ohio 45469

SYNOPSIS

Work in our laboratory has been concerned with the use
of acetylene chemistry to aid the processing of aromatic
heterocyclic polymers. Acetylene terminated heterocyclic
oligomers homopolymerize by both inter- and intramolecular
addition reactions leading to fused aromatic ring systems
which exhibit thermal and thermo-oxidative stabilities comparable with the heterocyclic backbone structures. The use
of pendant acetylene groups has demonstrated the potential
of a new curing mechanism providing a means to tougher
resins with use temperatures far exceeding processing temperatures. This paper will review the synthesis and characterization of these materials as well as some already
determined mechanical properties as composites and adhesives.

INTRODUCTION

High performance structural resins for advanced light-
weight, high temperature composite and adhesive applications
have led to the exploration of a number of polymeric systems.
The material requirements are not only for superior mechanical properties and durability, but also that the material
possesses environmental resistance and a high degree of

*Current Address: FRL, Dedham, Mass. 02026

thermo-oxidative stability. Environmental resistance is
primarily concerned with the effects of absorbed moisture
of the matrix or adhesive material. In the case of the
state-of-the-art epoxy resins, it has been shown[1-3] that
their elevated temperature properties are adversely affected
by high humidity to a degree that limits their ultimate
potential. The loss in properties with epoxide systems is
attributed to a plasticizing effect of the absorbed moisture
on the matrix resin.

The most significant problem in high temperature polymer applications is clearly one of processing. The characteristics which enable these polymers to perform at elevated
temperatures limit their solubility and fusibility, and thus
their processibility. The materials must exhibit moderate
glass transition temperatures, desirable solubility-fusibil-
ity characteristics and cure via addition reactions, all of
which would facilitate their fabrication as adhesives and
fiber reinforced composites.

One of the most promising classes of candidate materials for high temperature applications has been the aromatic
heterocyclic polymers. Unfortunately, the most thermally
stable systems in this class of materials are formed by condensation reactions with evolution of volatile by-products.
The volatile by-products which are evolved during fabrication form voids which greatly weaken the structures. To
circumvent the water of condensation, the polymers are usually fabricated as preformed high molecular weight polymers.
However, the restricted mobility of long chain molecules in-
hibits the interchain curing reactions during fabrication,
thereby causing the material to remain thermoplastic.

Work in our laboratory in the past few years has been
concerned with the use of acetylene chemistry, both to aid
the processing of aromatic heterocyclic polymers and to provide such materials a method by which they could become tougher and more durable in structural applications. The most
attractive feature of the acetylenic carbon carbon triple bond
is its capability to undergo various ionic and free radical
addition reactions, leading to highly fused thermally stable
aromatic systems. This paper will review our work on acetylene containing aromatic heterocyclic polymers with respect
to synthesis and characterization, as well as some already
determined mechanical properties as composites and adhesives.

RESULTS AND DISCUSSION

Terminal Acetylenes

The span of basic research of the 1960-70's identified a host of aromatic heterocyclic polymers which would fulfill the high temperature requirements for advanced structural applications. Research in the 1970's has been directed toward structural modifications of high temperature polymers to improve processibility without significantly altering the chemical structure which imparts high performance characteristics at elevated temperatures. The approach we have taken in our laboratory is to utilize reactive oligomers, where good solubility and fusibility properties are maximized. Acetylene terminated heterocyclic oligomers homopolymerize by both inter- and intramolecular addition reactions leading to fused aromatic ring systems which exhibit thermal and thermo-oxidative stabilities comparable with the heterocyclic backbone structures. The two heterocyclic systems which have been investigated are the polyphenylquinoxalines and aromatic polyimides.

The acetylene terminated quinoxaline oligomers, referred to as ATQ's, are prepared[4,5] from the reaction of benzil endcapped phenylquinoxaline oligomers and 3(3,4-diaminophenoxy)-phenylacetylene. The benzil end-capped oligomers are pre-

pared by the dropwise addition of a solution of diaminobenzi-
dine in m-cresol to a solution containing a two-fold excess
of the appropriate aromatic bis-benzil dissolved in m-cresol
containing a trace of acetic acid. Structural variation and
tailoring of the oligomers for specific processing parameters was accomplished by varying the aromatic bis-benzils.

The ATQ oligomers are soluble (30-35%) in low boiling
organic solvents such as methylene chloride, chloroform,
tetrahydrofuran and dioxane. Thermal analytical data for
the oligomers are shown in Table I. Analysis by DSC
($\Delta = 20°$/min) shows an initial strong baseline shift attributed to the softening point of the oligomers, while a strong
exotherm initiating at 200°C and maximizing at approximately
275°C for the polymerization of the terminal acetylene groups.
The extrapolated onset of the DSC baseline shift was taken
as the Tg. For the determination of glass transition temperature after cure, the TMA expansion (change of rate of
expansion) and for penetration modes (point of highest rate
of penetration) were used, also at a rate of 20°C/min.

TABLE I
Thermal analytical data of ATQ systems.

Ar	OLIGOMER	a Tsoft °C	b Tp °C	c Tg °C
⟨benzene⟩-O-⟨benzene⟩	ATQ-O	159	274	321
⟨benzene⟩-S-⟨benzene⟩	ATQ-S	144	277	331
⟨benzene⟩ (para)	ATQ-PP	161	277	340
⟨benzene⟩ (meta)	ATQ-MP	159	260	318

a. Tsoft = SOFTENING POINT OF UNCURED OLIGOMER (DSC)
b. Tp = TEMPERATURE AT WHICH THE POLYMERIZATION
           EXOTHERM IS AT A MAXIMUM (DSC)
c. Tg = GLASS TRANSITION TEMPERATURE OF CURED
           (8 HR AT 280°C) POLYMER (TMA)

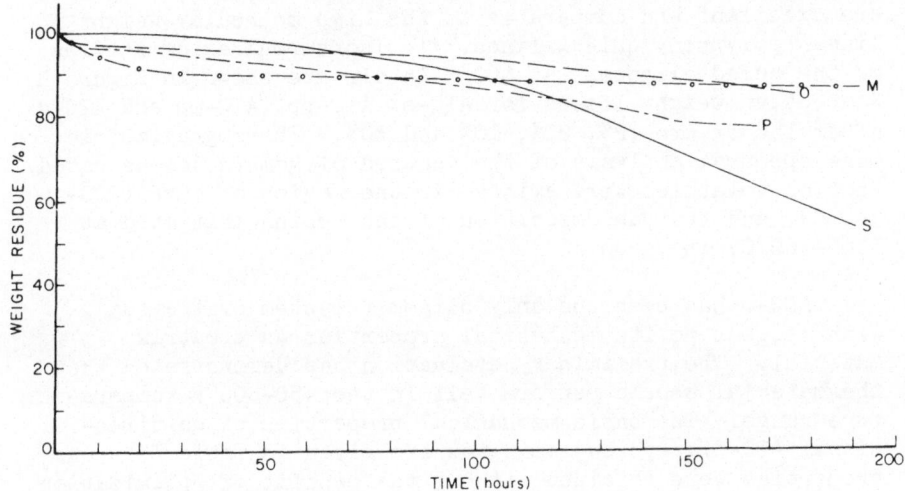

Fig. 1   Isothermal Aging In Air Of Cured ATQ's At 600°F

TABLE II

Properties of unidirectional ATQ-0/HTS composites.		
PROPERTY	R.T.	500°F
FIBER VOLUME	65 %	
FLEXURAL STRENGTH (Ksi)	221	194(88%)
FLEXURAL MODULUS (Msi)	19.5	18.1(93%)
SHORT BEAM SHEAR (Psi)	15,000	—
160°F – 30 DAY HUMIDITY EXPOSURE 40% EQUILIBRIUM MOISTURE GAIN		
	300°F	450°F
FLEXURAL STRENGTH (Ksi)	221(100%)	198(90%)
FLEXURAL MODULUS (Msi)	19.5(100%)	19.1 (98%)

The thermal-oxidative properties of the cured oligomers are excellent and comparable to the high molecular weight linear polyphenylquinoxalines.[6,7] Isothermal aging studies of the cured oligomers at 600°F in air are shown in Figure 1. Respective weight losses for ATQ-o, ATO-pp, ATQ-mp and ATQ-s after 150 hr are 12%, 22%, 10% and 30%. Thermogravimetric mass spectral analysis of the uncured oligomers demonstrated that no volatiles were emitted in the region of cure (200-300°C), and that decomposition of the resins initiated at (465-500°C) in vacuo.

ATQ-o has been the only oligomer system evaluated[8] with respect to its mechanical properties as a matrix material. The preliminary evaluation has demonstrated that the material should perform well in the 450-500°F temperature range. The basic mechanical properties of unidirectional ATQ-o/graphite laminates are given in Table II. The properties were obtained without the benefit of optimization of process parameters and a more comprehensive evaluation will be performed. Prepreg tapes have been prepared by drum winding graphite fiber bundles through (30-35%) solutions of the oligomer in various solvents such as dioxane, methylene-chloride and chloroform. The laminates were fabricated at 550°F for 2 hours under 200 PSI and then post-cured in a circulating air oven at 600°F for 16 hours.

The key objective of this work was to synthesize a thermoset quinoxaline system that would retain a high percentage of mechanical properties after adverse moisture conditioning. The quinoxaline backbone, as a thermoplastic system, has been shown to exhibit excellent mechanical property retention after long term high temperature, high humidity moisture exposure. What was not known in the ATQ system was the effect of moisture on the product of the addition reaction. Consequently, the composites were subjected to humidity aging tests by exposing flexural specimens to 160°F at 95% R.H. environment. After exposure, the property determinations were performed. After a thirty day moisture exposure, the ATQ-o thermoset system gained only 0.4% equilibrium moisture content and exhibited excellent high temperature mechanical property retention (Table II).

A more developed acetylene terminated heterocyclic system is the aromatic imide system, consisting of two oligomers, HR-600 and HR-650. The materials are synthesized[10] from benzophenonetetracarboxylic dianhydride and 1,3-bis(3-aminophenoxy)benzene. They differ in that the HR-600 oligomer is terminated with 3-aminophenylacetylene, whereas the HR-650 is terminated with 3-(3-aminophenoxy)phenyacetylene.

The oligomers are fully cyclized imide structures which are soluble in m-cresol, dimethylformamide and N-methyl-pyrrolidinone. The latter solvent is preferred for fabrication into fiber reinforced structural composites.

The HR-600 oligomer has been fabricated[11] into graphite composites with flexural strengths over 200 KSI, flexural moduli over 16 MSI and short beam shear strengths in the 17 KSI range. The material exhibits good mechanical property retention after 600°F aging in air. As an adhesive the HR-650 oligomer exhibits[12] ambient temperature titanium lap shear strengths of 3500-4000 PSI with good strength retention at 500°F.

The actual mechanism of the thermally induced acetylene reaction has not been determined and was first proposed[10] as a simple acetylenic trimerization. It is our opinion, however, that the terminal acetylene groups can simultaneously react via a number of alternate routes in addition to trimerization. Reaction of phenylacetylene and the model com-

pound, 2,3-diphenyl-6-(3-ethynylphenoxy)quinoxaline, under
the conditions of cure of the oligomers led[15] primarily to
highly fused polymeric materials. There were isolated, however, small quantities of complex mixtures of products.
Mass spectral analysis of these materials indicated the presence of triphenylbenzenes and phenylnaphthalenes. As a
comparison the reaction of phenylacetylene as a 40% solution
in benzene or methanol under high pressure at 220°C was recently reported[13] to give (75%) polymers, (5%) phenylnaphthalenes, and 20% triphenyl benzene isomers. The thermally
induced reaction of the terminal acetylene groups is very
complex and will require a significant research effort for
the exact mechanistic determination.

Pendant Acetylene Groups

     The state-of-the-art method of fabricating requires
flow of linear or branched resin molecules in order to bring
about wetting of the fiber reinforcement and molding. The
mobility of the molecules is then arrested during the curing
step in which reactive sites along the molecules are caused
to react with one another to form intermolecular bonds. The
result is generally a hardened, three-dimensional network
structure. However, an interesting paradox results, in that
while cure is meant to arrest molecular mobility, this
method of cure is itself very dependent upon molecular mobility to reach completion. When molecular movement ceases,
unreacted cross-linking sites become "frozen-in" and curing
stops. Raising the curing temperature will increase molecular movement and more reactive sites are thus "unfrozen",
allowing the curing process to continue. The implication
is the ultimate use temperature of the resin may only be as
high as the curing temperature used, since the resin may
undergo further softening at higher use temperatures. Aromatic heterocyclic resins are much more susceptible to this
mobility dependence than aliphatic resins due to their
inherent rigidity.

     We have employed pendant acetylene groups to demon-
strate a new cure concept referred to as the intramolecular
cycloaddition or IMC cure which is not dependent upon extensive molecular mobility. This consists of starting with
linear mobile polymer chains which display flow and mold-
ability at relatively low temperatures. The molecular

mobility is subsequently arrested by having pairs of pendant groups along the polymer chain undergo an intramolecular cycloaddition reaction to form a more rigid and therefore higher Tg structure.

The type of chemistry required to test the concept
is pairs of neighboring pendant groups which will undergo
a flow-independent, volatiles-free, intramolecular cycloaddition, through which a single bond is replaced by a
fused aromatic structure. A literature reaction which was
adaptable to our purpose is the thermal conversion of
2,2'-bis(phenylethynyl)biphenyl to 9-phenyl-dibenz[a,c]-
anthracene.[14,15] This reaction meets all the criteria of
an intramolecular cycloaddition and furthermore proceeded
in 80-100% yield.

The acetylenic biphenyl moiety was incorporated into
the polymer backbone of two aromatic heterocyclic systems,
the polyphenylquinoxaline and aromatic polyimide. The
introduction was carried out via the amino monomers,
2,2'-bis(phenylethynyl)-4,4',5,5'-tetraminobiphenyl and
2,2'-bis(phenylethynyl)-5,5'-diaminobiphenyl. The IMC type
of reaction depends only upon a rotational movement of the
polymer backbone which requires substantially less molecular mobility than the translational movement needed for the
intermolecular cure. Therefore, the curing reaction can
continue to completion long after the resin is essentially
vitrified, and the resultant use temperature should be
substantially higher than the cure temperature.

Condensation of the phenylethynyl-tetraamine with
m-bis(p'-phenoxyphenylglyoxalyl)benzene in m-cresol,
afforded prepolymer A. The prepolymer exhibited[16,17] a
Tg of 215°C by TMA and a strong DTA exotherm maximizing
at 245°C. Thermal treatment of the prepolymer at 245°C
overnight resulted in its conversion to the fused benzan-
thracene polymer B as evidenced by absence of the DTA exotherm and the acetylene absorption band in the infrared.
Polymer B, after the 245°C cure, showed a Tg of 365°C
which represents a 120°C advancement in Tg over the cure
temperature.

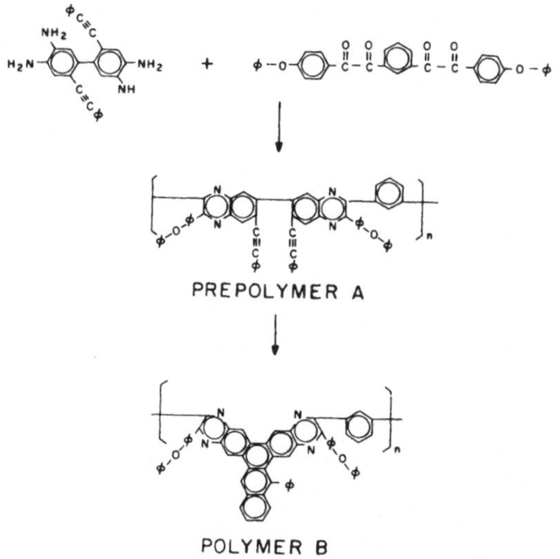

PREPOLYMER A

POLYMER B

The aromatic polyimide system was prepared by the
reaction of the diaminoethynylbiphenyl moiety with 2,2'-
bis[4-(3,4-dicarboxyphenoxy)phenyl] hexafluoropropane
dianhydride. The polymerization was carried out in
N,N-dimethylacetamide to form the polyamicacid and then
cyclohydrated to the imide by addition of acetic anhydride.
Although several anhydrides[18] were condensed with the
diamine, the anhydride containing the hexafluoropropane
group provided the lowest Tg with the rod-like imide
system.

PREPOLYMER C

POLYMER D

The imide prepolymer (C) exhibited a Tg of 185°C by TMA. Analysis of the material by DSC (20°C/min) showed the IMC reaction exotherm began around 165°C but did not show a significant rate of increase until above 200°C with a maximum at 233°C. The Tg of the cured polymer D, after the IMC reaction (treatment at 233°C overnight), appeared at 310°C by TMA and 325°C by DSC.

The prepolymers (A) and (C) have not as yet been fabricated into specimens for high temperature mechanical property evaluation. In theory, both materials could be fabricated at the softening range of 200-225°C, cured at 250°C, and the cured materials would be usable at temperatures up to 300-350°C. In practice, the materials suffer from a problem which has been found to be common to all high molecular weight thermoplastic polymers, specifically that the extent of flow at the Tg is insufficient for satisfactory processing.

CONCLUSIONS

Acetylene terminated aromatic heterocyclic oligomers possess excellent processing characterisitics for 450-600°F matrix resins. The terminal acetylene groups homopolymerize by both inter- and intramolecular addition reactions leading to highly fused aromatic ring systems which are moisture insensitive and exhibit excellent thermal oxidative stability and mechanical properties. Follow-on research in this area is directed toward determining the exact mechanism of the cure reaction in order to be able to control the polymer network or chain structure.

The use of acetylene groups pendant to heterocyclic polymers has demonstrated a new cure concept (IMC) which should provide tougher, more durable matrix resins with use temperatures far exceeding cure temperatures. Current and future work in this area is directed toward solving the problems associated with the extent of flow at the softening points of the resin systems.

REFERENCES

1. C.E. Browning and J.M. Whitney, Amer. Chem. Soc., Div. Org. Coatings and Plastics Chem. Prepr., 33 (2)(1973).

2. C.E. Browning and J.M. Whitney, Amer. Chem. Soc. Advances in Chemistry Series, 137 (1974).

3. C.E. Browning and J.T. Hartness, Composite Materials: Testing and Design (Third Conference), ASTM STP 546, 284 (1974).

4. R.F. Kovar, G.F.L. Ehlers and F.E. Arnold, Amer. Chem. Soc., Div. Polym. Chem. Prepr., 16 (2), 247 (1975).

5. R.F. Kovar, G.F.L. Ehlers and F.E. Arnold, J. Polym. Sci., Polym. Chem. Ed. (in press).

6. J.M. Augl, J.V. Duffy and S.E. Wentworth, J. Polym. Sci., Polym. Chem. Ed., 5, 1023 (1974).

7. P.M. Hergenrother, Amer. Chem. Soc., Div. Org. Coatings and Plastics Chem. Prepr., 35 (5), 166 (1975).

8. A. Wereta Jr., T. Hartness, C.E. Browning and R.F.
    Kovar, Soc. Adv. Mat'r. and Process Eng. Series 21,
    83 (1976).

9. R.F. Kovar and F.E. Arnold, Soc. Adv. Mat'r. and
    Process Eng. Tech. Conf. 8, 106 (1976).

10. A.L. Landis, N. Bilow, R.H. Boschan, R.E. Lawrence
    and T.J. Aponyi, Am. Chem. Soc., Div. Polym. Chem.
    Prepr. 15 (2), 537 (1974).

11. N. Bilow, A.L. Landis and T.J. Aponyi, Soc. Adv.
    Mat'r. and Process Eng. Series 20, 618 (1975).

12. R.H. Boschan, A.L. Landis and T.J. Aponyi, Soc. Adv.
    Mat'r. and Process Eng. Series 21, 356 (1976).

13. W. Jarre, D. Bieniek and F. Korte, Naturwissen-
    schaften, 62 (8), 391 (1975).

14. S.A. Kandil and R.E. Dessy, J. Am. Chem. Soc., 88,
    3027 (1966).

15. E.H. White and A.A.F. Sieber, Tetrahedron Letters,
    2713 (1967).

16. F.L. Hedberg and F.E. Arnold, Am. Chem. Soc., Div.
    Polym. Chem. Prepr., 16 (1), 677 (1975).

17. F.L. Hedberg and F.E. Arnold, J. Polym. Sci., Polym.
    Chem. Ed. (in press).

18. F.L. Hedberg and F.E. Arnold, AFML-TR-76-198 (1976).

DISCUSSION

J. K. STILLE - U. OF IOWA. With regard to your intramolecular cyclization reaction, once that reaction is
finished, once that cure has taken place, were your polymers
still completely soluble?   F. E. ARNOLD - AIR FORCE
MATERIALS LABORATORY. No, they weren't. There is some
intermolecular reaction, the extent of which we do not know
because we couldn't fully characterize them.

R. NATARAJAN - LORD CORP. The interlaminar shear
strength that you got with the ATQ's and graphite fiber is
quite good being of the order of 15,000 psi. Have you
tried this resin system with Kevlar® aramid fiber where
you have a serious interlaminar shear strength problem?
F. E. ARNOLD. No, we haven't fabricated these systems with
the Kevlar® aramid fibers as yet. We do plan to do this to
see what effect it has on the modulus and creep properties.

BIODEGRADABLE POLYMERS FOR SUSTAINED DRUG DELIVERY

A. Schindler, R. Jeffcoat, G. L. Kimmel,
C. G. Pitt, M. E. Wall, and R. Zweidinger

Research Triangle Institute
P. O. Box 12194
Research Triangle Park, N.C.   27709

SYNOPSIS

Most studies of the sustained release of drugs from subdermally implanted polymer devices have centered on the use of silicone rubber. However, the use of this polymer has serious limitations because of its non-biodegradability. The depleted capsule has to be surgically removed if one is to eliminate potential problems associated with non-degradable foreign substances remaining in the body for an indefinite length of time. The development of polymer systems which combine the release properties of silicone rubber with biodegradability will represent a significant advance in the technique of controlled release of contraceptives. The polymer system selected for our investigations were polyesters comprising homo and copolymers of glycolide, dilactide, ε-caprolactone, and ε-decalactone. These polymers were found to undergo random hydrolytic degradation under in vitro and in vivo conditions. The polymers were characterized by their rates of biodegradation and their release parameters for contraceptive steroids. Long time release rates of steroids from monolithic and reservoir devices were determined. Especially poly(ε-caprolactone) was found to come close to meeting the requirements of a biodegradable reservoir device for controlled drug delivery, with a useful lifespan approaching one year. Copolymers of ε-caprolactone and racemic dilactide are more permeable than poly(ε-caprolactone) and are of value for biodegradable devices with shorter than one year lifespan.

INTRODUCTION

Until recent times the routes of drug administration have remained essentially unchanged, consisting almost exclusively of either oral or parenteral administration. The necessity for new techniques becomes apparent with the development of drugs with increased potency. Increased potency with minimal persistence or side-effects is usually a consequence of rapid metabolism which means effectiveness only within a narrow limit of time and concentration. Repeated application of a drug in individual doses generates strongly fluctuating drug levels in the body with the possibility of overdose or underdose.

Provided the drug is continuously delivered at a constant rate by a controlled-release device and its removal follows first order kinetics then a stationary drug level will be established given by the ratio of both rate constants. The stationary level can be kept extremely low if the delivery device is placed close to the target organ.

A controlled-release delivery system is a combination of a biologically active agent (drug) with an excipient, commonly a polymeric material which can play either an active or a passive role in the delivery process. In the first case the drug is released from the polymeric matrix by changes in the chemical or physical properties of the latter. Such changes can involve, among others, biodegradation of the polymer (surface errosion), progressive swelling with subsequent drug diffusion from the swollen region, and hydrolysis of drug-polymer bonds.

The polymer will play a passive role if it acts solely as a barrier which controls the rate of drug delivery by diffusion. Indeed, changes in the properties of the polymer are undesirable in this case since thereby the parameters governing the diffusion process will change. Purely diffusion controlled delivery systems generally belong to either one of two types, monolithic devices or reservoir devices.

In monolithic devices the drug is uniformly mixed with the polymeric matrix and is present either in dissolved or dispersed form. For a dissolved drug

Fick's laws apply for the release kinetics whereas the
release of dispersed drugs can be described by Higuchi's
equation (1). Release kinetics for both conditions and
covering different device geometries were recently
reviewed by Baker and Lonsdale (2).

In reservoir devices the drug is enclosed within an
inert polymer membrane. As long as the thermodynamic
activity of the drug within the device does not change
(saturated solution with excess solid drug) the concentration gradient across the polymer membrane remains
constant provided there is no concentration change on
the exit side of the membrane. During this time the
drug will be released at a constant rate (zero order
release). The release kinetics will change to first
order as soon as the last solid drug particle dissolves.

Reservoir devices are the only ones providing zero-
order release for extended time periods. The release
rates for monolithic devices containing large amounts of
dispersed drug fall according to a $(time)^{1/2}$ law essentially throughout their life span. With dissolved drug
the $(time)^{1/2}$ law applies with good approximation for
about the first half of the device life; subsequently,
the release rate decreases exponentially (2).

Most studies of the controlled release of drugs
from subdermally implanted polymer devices have centered
on the use of silicone rubber (Silastic). In 1964
Folkman and Long (3) demonstrated the potential of this
polymer for the controlled release of certain cardioactive
steroids. The good permeability of silicone rubber to a
variety of drugs and its favorable biocompatibility en-
couraged other workers in its use as membrane material
for diffusion controlled delivery devices (4).

Despite good permeability properties for a large
number of drugs and facile processibility the application
of silicone rubber as implant material has serious
limitations because of nonbiodegradability. The depleted
capsules must be surgically removed if one is to eliminate potential health hazards associated with nonde-
grading foreign substances remaining in the body for an
indefinite length of time. Particularly in areas or
countries where proper medical care is not readily available,
surgery, even of minor size, might present a true problem.

A significant advance in the technique of controlled release will be given by the development of polymer systems which combine the release properties of silicone rubbers with biodegradability. It is immediately obvious that the rate of biodegradation has to be coordinated with the projected life span of the device, i.e., during the release period the release properties of the device should remain essentially unchanged. In the ideal case substantial degradation should commence after depletion of the device.

Biodegradable implants for controlled delivery of narcotic antagonists were first described by Yolles et al. (5). The delivery devices were either drug loaded films with or without plastization by tributyl citrate or drug loaded polymer powders suspended in a dispersing agent (5-8). In continuation of this work the application of poly(L-lactide) was extended to a number of different drugs (9,10). Poly(L-lactide) and racemic poly(lactide) were later used for the controlled delivery of contraceptives in form of monolithic devices (11) or as reservoir devices prepared by microencapsulation techniques (12,13) for the delivery of narcotic antagonists.

A polymer which is to be used in a biodegradable drug delivery system must be tailored to meet a number of specific requirements. If tissue irritation is to be minimized the polymer should be soft and pliable. This means little or no crystallinity and a glass transition temperature not higher than body temperature. The use of plasticizer is not suitable for long time applications because of its uncontrollable loss with concomittant change in permeation properties. The polymer must be compatible with the drug (i.e., nonreactive) and the body (i.e., nontoxic). The latter condition includes all degradation products as well as their possible metabolites. Most importantly, the permeability to the drug and the rate of biodegradation should be compatible with the application in mind. For the controlled release of contraceptives a minimum usable life of 6 months but preferable of 12 months will be required. As a final condition the polymer should be obtainable from inexpensive, commercially available starting materials by a simple polymerization process.

All these conditions put a severe constraint on the selection of a suitable polymer system. The system selected for our investigations comprised poly(lactones) with emphasis on homo- and copolymers of racemic dilactide, ε-caprolactone, and ε-decalactone.

LACTONE POLYMERIZATION

The lactones selected by us can be polymerized by a ring-opening mechanism initiated by a wide range of different catalysts (14-18). In order to avoid catalysts possessing metal-carbon bonds all our polymerizations were performed with either stannous octoate or stannous chloride in a concentration range from 50 to 500 ppm. Polymerizations were generally in bulk at 130°C. In all polymer preparations the usual high vacuum technique was applied.

Poly(Lactide)

Dilactide, the cylic diester of lactic acid, possesses two asymmetric carbons and exists in three configurations: the two enantiomers D- and L- dilactide melting at 95°C, and the meso-form melting at 41°C. Both enantiomers form a racemic compound (DL-dilactide) melting at 127°C (19).

In polymerizations catalyzed with tin salts ring-opening occurs exclusively by acyl cleavage with retention of the optical configuration of both asymmetric carbons (15). Poly(L-lactide) is isotactic (all m-dyads) whereas poly(DL-lactide) resulting from racemic dilactide can be considered atactic although there exist some restrictions in the number of possible stereosequences. In each growth step two lactide units are added to the growing chain which form a m-dyad. Because of this peculiarity of the propagation step at least half of the dyads have to be meso and the racemic dyads must always be separated by an odd number of m-dyads. Analysis of ^{1}H-NMR spectra of poly(DL-lactide) revealed Bernoullian statistics to apply (20).

Polymerizations initiated by nucleophiles such as tertiary amines proceed via macrozwitterions with ring opening at the oxygen-carbon (sp^3) bond. With this mode

of ring-opening inversion of the configuration of the
participating carbon can occur. The polymerization seems
to be complicated by reformation of dilactide by "back-
biting" at the growing chain end. This possibility is indicated by a nearly total loss in optical activity of a
polymer derived from L-dilactide and by the presence of
meso-dilactide in the residual monomer.

Poly(L-lactide) is highly crystalline with a melting
point T_m = 180°C and a glass transition temperature
around 67°C. Poly(DL-lactide) is amorphous and has a
glass transition temperature around 57°C.

With tin catalysts and proper polymerization conditions, poly(lactide) with number average molecular
weights up to about 4×10^5 ([η] in chloroform about 7.3
dℓ/g) can be obtained. These molecular weights are
still below those calculated from catalyst concentrations
assuming living chain ends (15). The discrepancy is
very likely due to the presence of traces of lactoyllactic acid which acts as a chain transfer agent. The
polymerization is very sensitive to the presence of
hydroxyl groups and, indeed, high molecular weight
polymers were obtained only with the use of silanized
glass vessels.

Poly(DL-lactide) is soluble in most common organic
solvents such as benzene, acetone, THF, and chlorinated
hydrocarbons. Poly(L-lactide) is soluble in chloroform
and methylene chloride and in benzene at elevated
temperature (50-60°C).

The shelf life of poly(lactide) is good although a
slight decrease in molecular weight with time can be
observed depending on the purity of the polymer. Improp-
erly purified polymer, i.e., containing substantial
amounts of residual dilactide, degrades relatively fast
due to the facile formation of lactoyllactic acid which
catalyzes further polymer degradation.

Carboxyl end group contents of poly(lactide) were
determined by a dye interaction technique proposed by
Palit and Mandal (21). A dilute solution of the polymer
in benzene is reacted with a dilute benzene solution of
a basic dye (e.g., Rhodamine 6G) in its unionized basic

form and the resulting color change at 515 nm is evaluated spectrophotometrically. Spectrophotometric measurements are calibrated with polymers possessing sufficiently high carboxyl group contents to be determined by potentionmetric titration in acetone.

The carboxyl group content of the polymer, [COOH] in mole/g, was then found to be given by

$$[COOH] = 1.77 \times 10^{-5} \, (\Delta E / C_p) \tag{1}$$

where ΔE is the change in optical density at 515 nm, and C_p is the polymer concentration in g/l. The method is rather sensitive, and permits end group determinations of polymers with molecular weights of several hundred thousands.

The dye interaction technique was used to follow polymer degradations in in vivo and in vitro studies as well as to establish a viscosity-molecular weight relationship for racemic poly(lactide). In order to assure the presence of one carboxyl group per chain partially hydrolized polymers of high initial molecular weights were used in the latter case. The polymers were hydrolyzed with deionized water either in form of solid powder or in tetrahydrofuran or acetone solution with complete agreement of data obtained in both ways.

A plot of number average molecular weights from carboxyl group determinations versus intrinsic viscosities is shown in Fig. 1. The corresponding Mark-Houwink equation for [η] in dl/g is given by

$$[\eta] = 3.64 \times 10^{-4} \, M_n^{0.75} \tag{2}$$

The data of Fig. 1 also include some representing low molecular weight polymer prepared in the presence of transfer agents such as water or lactoyllactic acid. In this case good agreement with partially hydrolyzed polymer is obtained. High molecular weight polymers generally possess less carboxyl groups than corresponding to the molecular weight resulting from equ. (2). This finding could mean either a molecular weight distribution considerably narrower than that of partially hydrolyzed samples or more likely the presence of end

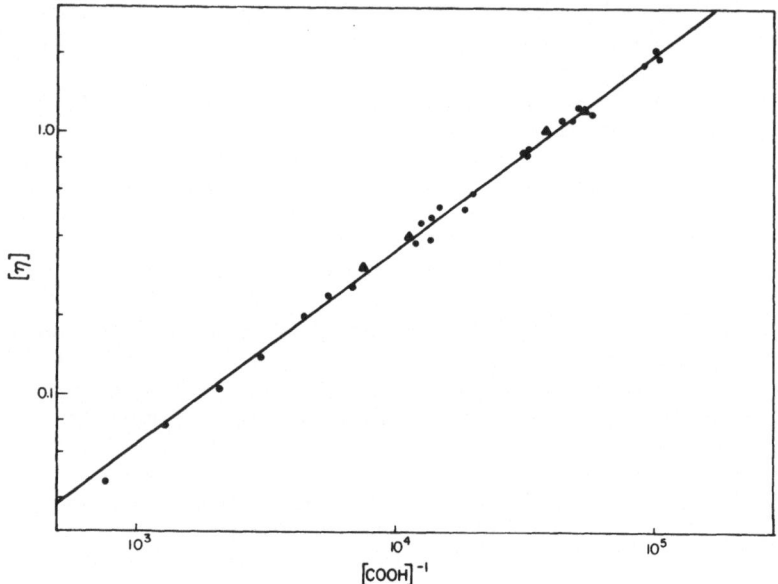

Fig. 1 – Relationship between intrinsic viscosity
in benzene and carboxyl content of partially
hydrolyzed racemic poly(lactide).

groups other than carboxyl, e.g., catalyst residues.
The presence of carboxyl end groups can be almost completely suppressed with the use of alcohols as transfer
agents.

Poly(lactide) is rather sensitive to γ-radiation
and is severely degraded even at low doses. This behavior has to be considered if sterilization of implants
by γ-irradiation is intended.

Poly(ε-Caprolactone)

Polymerization of ε-caprolactone and polymer properties were reviewed in detail by Brode and Koleske
(16). Our polymerizations were generally performed in
bulk with stannous octoate (200–1000 ppm) at 130°C.
Intrinsic viscosities in benzene of the resulting polymers
were usually between 1.2 and 2.3 dl/g, higher values
being difficult to obtain due to trace impurities in the
monomer.

Gel permeation chromatography on three polymer samples with intrinsic viscosities in benzene of 1.12, 2.01, and 3.53 dl/g revealed M_w/M_n ratios of 3.3, 3.5, and 2.7, respectively. These values agree with those generally found for dibutylzinc-catalyzed polymerizations (22).

Poly(caprolactone) is a crystalline polymer with a melting point at 63°C and a glass transition at -60 to -70°C. The polymer crystallizes very readily and connot be quenched to a glass (16).

The ambient temperature stability (shelf life) of poly(caprolactone) is strongly affected by its carboxyl content (16). Upon storage at ambient temperature the polymers display a gradual decrease in molecular weight with time which seems to proceed autocatalytically and is related to a random hydrolytic reaction. The average decrease in intrinsic viscosity observed for several samples after 12 and 24 months of storage was 10 and 25%, respectively.

Thermal degradation of poly(caprolactone) at 220°C was recently investigated (23). It proceeds by a zipping-off mechanism yielding monomeric caprolactone without change in reduced viscosity up to a weight loss of 53%. This means that the rate of random dissociation is small as compared to the rate of depropagation and an active species once formed will disappear immediately.

Carboxyl end group determination by the dye method was not successful with poly(caprolactone) due to the high pK_a of the carboxyl groups which can be assumed to be similar to that of hexanoic acid (pK_a = 4.89). The pK_a of carboxyl groups in poly(lactide) will be considerably lower and very likely close to that of lactoyl lactic acid, pK_a ~ 3.0 (24).

From sol-gel analysis (25) of γ-irradiated poly-(caprolactone) the ratio of chain scission to cross-linking was determined to be about 0.9, as compared with the corresponding value of 0.3-0.4 for polyethylene. The critical dose for a polymer sample of intrinsic viscosity 1.9 dl/g was about 4 megarads.

Copolymers

Permeability determinations with progesterone
yielded a permeability constant for poly(caprolactone)
($P = 8 \times 10^{-6}$ 0cm^2/sec) rather similar to that of
Silastic (26) ($P = 1.6 \times 10^{-5}$ cm^2/sec) whereas racemic
poly(lactide) exhibited a permeability about five orders
of magnitude lower ($P = 1.8 \times 10^{-10}$ cm^2/sec). Consequently, further studies concentrated on poly(caprolactone).

It could be anticipated that poly(caprolactone)
containing small amounts of a comonomer would have
improved use properties. The decrease in polymer crystallinity would improve both the flexibility and the drug
permeability of the delivery device. Furthermore, by
selecting different comonomers the rate of biodegradation could be adjusted. Two comonomers were investigated in greater detail, dilactide and ε-decalactone.

__Dilactide-co-ε-Caprolactone.__ Copolymerizations of
dilactide and caprolactone were performed with stannous
octoate, stannous chloride, and tetrabutyl titanate
(TBT) as catalysts at 130 and 180°C either in bulk or in
the presence of toluene. Copolymer compositions were
determined by ^{1}H-NMR spectroscopy, specific optical
rotation in the case of L-dilactide, or by ^{3}H-contents
using ^{3}H-labeled caprolactone. Agreement of the three
analytical methods was generally good although ^{1}H-NMR
spectroscopy soon reached its limit of applicability at
lower lactide concentrations.

Copolymerization parameters for different polymerization conditions are summarized in Table I. Under all
polymerization conditions the dilactide preferentially
polymerizes with itself and copolymers of rather uneven
composition result if the conversion exceeds a value of
30-40%. Since there are strong indications (15,22) that
the lifetimes of the growing chains are extremely long
the copolymers very likely possess a block structure
which might be randomized to some degree by transesteri-
fication reactions. While TBT exhibited the most favorable copolymerization parameters only relatively low
molecular weight products are formed ($[\eta] \sim 0.6$ with TBT
versus $1.5 - 2.0$ with stannous octoate). Instantaneous
copolymer compositions as a function of conversion are

Table I. Effect of Polymerization Conditions on the Co-Polymerization Parameters (r_1, r_2) of Dilactide and Caprolactone

Dilactide	Catalyst[a]	r_1 (Dilactide)	r_2 (Caprolactone)
DL or L	SnOct	34.7	0.24
DL or L	SnOct	29.0[b]	0.46[b]
L	SnCl$_2$	9.5[c]	0.30[c]
DL or L	SnCl$_2$	6.8	0.19
DL	TBT	5.2	0.43

[a] Sn-Oct = Stannous octoate; TBT = Tetrabutyl titanate.

[b] Toluene added, all other polymerizations in bulk.

[c] Polymerization temperature 180°C, all others 130°C.

shown in Fig. 2 for an initial monomer ratio of 1:1 catalyzed by stannous octoate and TBT.

Glass transition temperatures of copolymers as function of their lactide contents are shown in Fig. 3. The solid line corresponds to the Fox equ. (27)

$$1/T_{1,2} = (w_1/T_1) + (w_2/T_2) \qquad (3)$$

where $T_1 = 203°K$ is the T_g value for poly(caprolactone) and $T_2 = 330°K$ the T_g value for poly(lactide). $T_{1,2}$ represents the T_g value of a copolymer containing weight fractions w_1 and w_2 of caprolactone and lactide units, respectively. Copolymers with T_g below room temperature are very tough and slightly hazy rubbers of good elasticity, the necessary cross-links being supplied by the crystalline phase formed by longer caprolactone segments.

Although the mechanical properties of these co-polymers satisfied the requirements of a drug delivery device their stability during extended periods of storage at ambient temperature present some problems. Figure 4 shows the effect of copolymer composition on polymer degradation during storage for periods between 145 and 245 days.

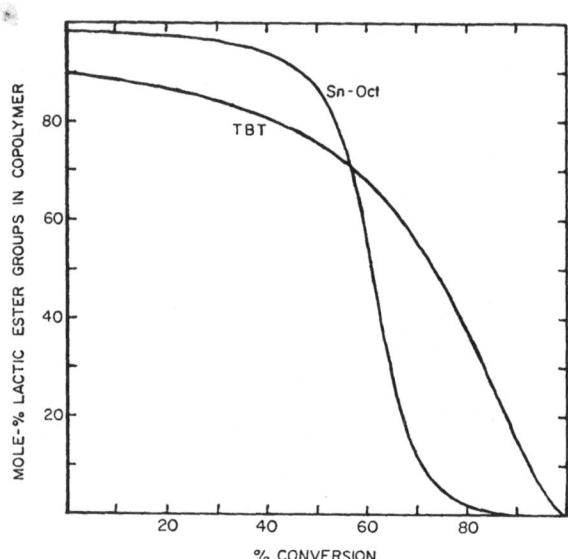

Fig. 2 – Effect of conversion on the instantaneous
         copolymer composition in dilactide-caprolac-
         tone copolymerization catalyzed with stannous
         octoate (Sn-Oct) and tetrabutyl titanate
         (TBT) for an initial monomer ratio of 1:1

Data of Fig. 4 reveal that the stability of the
copolymers decreases dramatically with increasing
lactide content until the glass transition temperature
exceeds ambient temperature. Copolymers containing L-
lactide units are more stable very likely due to partial
crystallization of longer lactide segments. Copolymers
kept at -20°C remained completely undegraded after
storage for over one year.

ε-Decalactone-co-ε-Caprolactone. Copolymers of ε-
decalactone and ε-caprolactone were prepared in bulk at
130°C with stannous octoate as catalyst. For accurate
determinations of copolymer compositions ^{3}H-labeled ε-
caprolactone was used. The copolymerization parameter
for caprolactone was found to be 2.2, whereas the parameter for ε-decalactone was practically zero. The
latter finding agrees with homopolymerization experiments
which established that ε-decalactone polymerizes extremely
slowly to low molecular weight products ([η] ~ 0.4 dl/g).

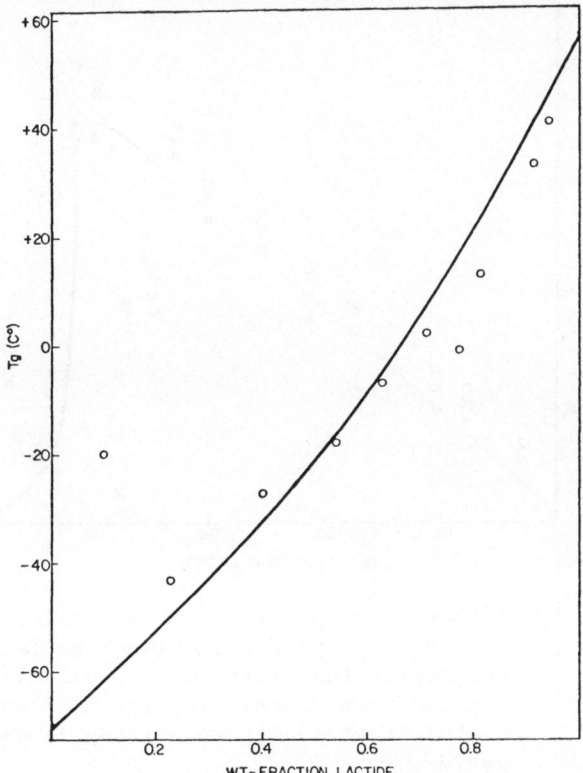

Fig. 3 - Effect of copolymer composition on glass
transition temperature for DL-dilactide-co-
ε-caprolactone

IN VITRO AND IN VIVO DEGRADATIONS

In vitro and in vivo degradation studies were
performed with polymer samples in film or tube form.
Polymer films were either cast from methylene chloride
solution or compression molded with thicknesses between
100-300 μm. Polymer tubes from poly(caprolactone) and
caprolactone copolymers were made by tightly wrapping
several layers of a polymer film around a Teflon support
(tubes with O.D. of 1.2 mm). The package was then
heated under rotation in vacuum until a polymer tube of
even wall thickness (150-600 μm) was formed. Before use
the tubes were heat-sealed on both ends.

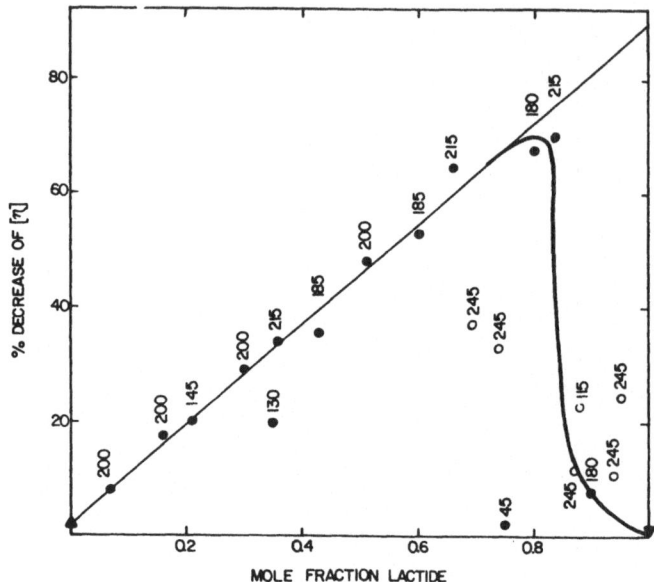

Fig. 4 – Effect of composition on the ambient tempera-
        ture stability of dilactide-caprolactone co-
        polymers. Full circles: DL-dilactide-co-
        caprolactone, open circles: L-dilactide-co-
        caprolactone. Storage periods in days indi-
        cated

    In vitro studies were performed in deionized water
or in buffer solutions under thermostated conditions.

    Female New Zealand white rabbits were used for in
vivo studies. The polymer samples (films: 2 x 1 cm,
tubes: ~2 cm long) were subcutaneously implanted via
incisions about the midline of the dorsal region. At
specified time intervals duplicate implants were removed
from different rabbits and visually inspected for evidence of biodegradation or tissue ingrowth. If no
degradation was apparent and the implant could be
cleanly excised, it was dried in vacuo, weighed, and
characterized by its intrinsic viscosity. The weight
loss of radiolabeled polymers was assayed by dissolution
in benzene and determination of the radioactivity of an
aliquot. With homopolymers of dilactide the extent of
degradation was also followed by carboxyl group determinations.

Racemic Poly(Lactide)

In vitro degradations of racemic poly(lactide) films in water at 60°C are shown in Fig. 5. The progress of degradation was followed by carboxyl group determinations and intrinsic viscosity measurements. Except for the initial part of the degradation curve, the carboxyl content of the polymer increases exponentially with time, i.e., the rate of carboxyl group formation (hydrolytic chain scission) is proportional to the concentration of carboxyl groups thus following a simple first order law

$$d[COOH]/dt = k[COOH] \qquad (4)$$

or
$$[COOH] = [COOH]_o \exp(kt) \qquad (5)$$

Initially, the rate of carboxyl group formation is somewhat higher than observed during the later part of degradation which possibly reflects the presence of a small percentage of hydrolytically more labile linkages.

The observed direct proportionality between rate of hydrolytic chain scission and carboxyl concentration is somewhat surprising since one might anticipate a square root dependence which expresses the dissociation of a weak acid. Such a dependence was found for the hydrolytic degradation of poly(ethylene terephthalate) (28).

If one neglects changes in form and width of the molecular weight distribution accompanying the random degradation process an exponential rate law should also apply for the change in intrinsic viscosity

$$[\eta] = [\eta]_o \exp(-\alpha kt) \qquad (6)$$

where α is the exponent in the Mark-Houwink equation

$$[\eta] = K\bar{M}_n^{\alpha} \qquad (7)$$

Viscosity data of Fig. 5 are in good agreement with equ. (6).

The value of the first order rate constant of hydrolysis at 60°C was found to be 0.56 ($\pm$0.04) day^{-1}. The corresponding value at 30°C is 1.2 ($\pm$0.2) x 10^{-2} day^{-1}.

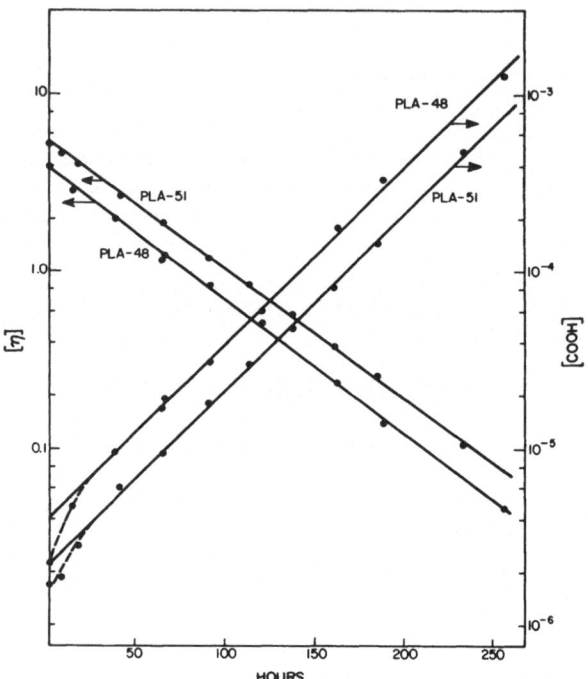

Fig. 5 – Change of carboxyl content and intrinsic
         viscosity during the hydrolysis of racemic
         poly(lactide) films at 60°C

        In vivo degradation studies were performed with
four polymer samples in film form possessing initial
viscosities of 0.47, 1.15, 2.24, and 3.24 dl/g corresponding to number average molecular weights of 15, 50,
120, and 200 x 10^3, respectively. Degradations were
again followed by carboxyl group determinations (Fig. 6)
and viscosity measurements (Fig. 7) on the excised
implants. Both presentations of the degradation process
show the same rate law to apply as found for in vitro
experiments. The carboxyl data are slightly more scattered than the viscosity data which is not surprising
since even very small tissue residues which are of
no effect in viscosity determinations will strongly
affect the sensitive dye test.

        No weight loss could be detected on the excised
implants as long as their intrinsic viscosity did not

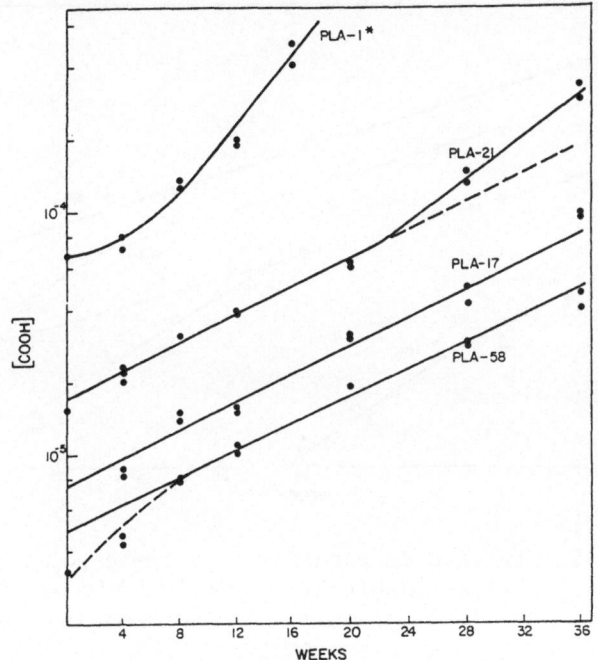

Fig. 6 - In vivo degradation of racemic poly(lactide)
films (rabbits). Change in carboxyl content
of excised implants

decrease below a value of about 0.5 dl/g. When this
value is approached the degradation accelerates and is
accompanied by severe weight losses.

The average rate constant for the initial part of
the biodegradation process ([η] > 0.5) was found to be
8.6 (+0.4) x 10^{-3} day^{-1} which means that it takes about
80 days for a poly(lactide) implant to degrade to half
its average molecular weight.

The absence of observable weight loss together with
the exponential increase in carboxyl end group content
implies an autoaccelerated, random degradation process
occurring throughout the bulk of the sample. The simul-
taneous occurrence of surface errosion cannot be ruled
out but its contribution to the overall process is too
small to become observable at the surface to volume
ratios of our implants.

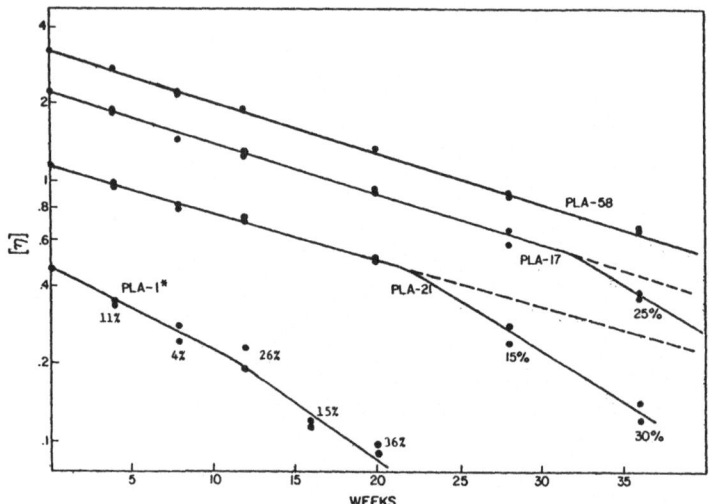

Fig. 7 - In vivo degradation of racemic poly(lactide)
films (rabbits).   Change in viscosity of
excised implants. Weight loss indicated.

Poly(ε-Caprolactone)

In vivo degradation studies with poly(caprolactone)
were performed with samples in film and tube form with
the initial intrinsic viscosities covering a range of
0.5-2.4 dl/g. The change in intrinsic viscosity of
excised implants with implantation time is shown in Fig. 8.

Again as in the case of poly(lactide) degradation
the experimental results can be described formally by
equ. (6). Independent of the initial molecular weight
all polymer samples degraded with an average rate constant of 3.0 ($\pm$0.1) x 10^{-3} day^{-1}.

As in the biodegradation of poly(lactide) no weight
loss was observed with poly(caprolactone) implants until
their viscosity decreased below a value of about 0.4 dl/g.

The good agreement between degradation data found
for the same polymer (PCL*-1) in film and tube form,
representing more than a tenfold change in surface to
volume ratio, indicates again a uniform degradation of
the bulk of the sample.

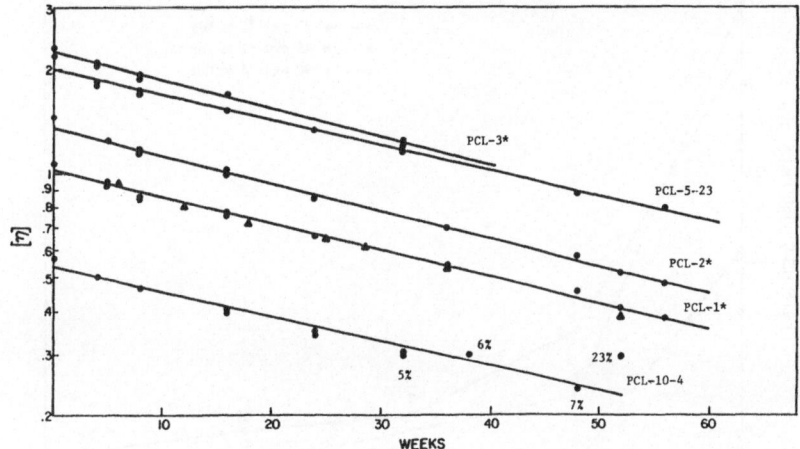

Fig. 8 - In vivo degradation of poly(caprolactone)
tubes (•) and films (▲) (rabbits). Change
in intrinsic viscosity of excised implants

Electron scanning micrographs of the surface of
excised implants revealed some surface errosion to take
place. A low molecular weight poly(caprolactone) sample
($[\eta]_0$ = 0.63) with extensive spherulite formation revealed
onset of errosion at the spherulite boundaries. With
time the resulting cavities increased in width whereby
individual spherulites or groups of spherulites became
isolated. During the degradation the samples showed a
continuously increasing weight loss whereas their
viscosities remained practically constant after a small
initial decrease.

With high molecular weight polymers only irregular
pit formation without distinct features is observed.

Copolymers

The low ambient temperature stability of copolymers
composed of dilactide and caprolactone indicated a fast
rate of biodegradation to be expected. Indeed, in vivo
studies with three copolymers containing 27, 43, and 90
mole-% lactide units revealed rapid biodegradation to
take place with 50% weight loss occurring after about 6,
12, and 16 weeks, respectively. As shown in Fig. 9 the

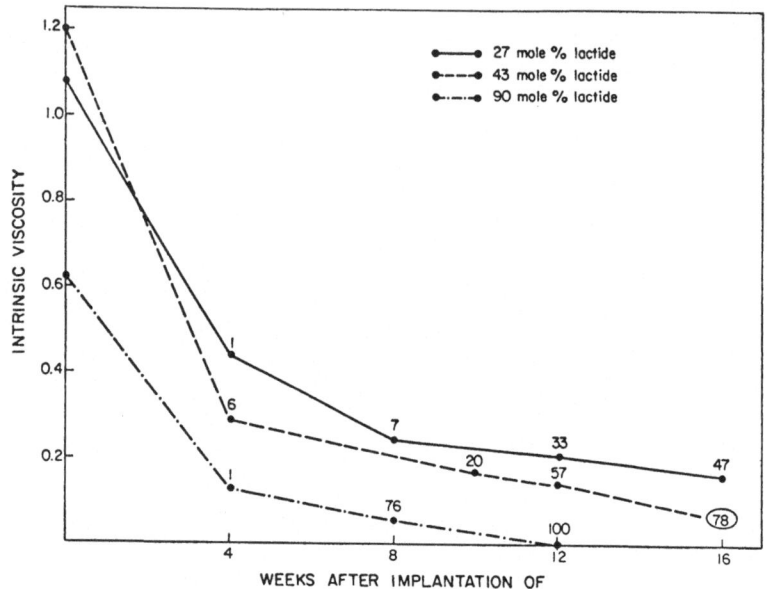

Fig. 9 - In vivo degradation of dilactide-ε-caprolac-
         tone (LA/CL) copolymers in tube form. Per-
         cent weight loss indicated

molecular weights of the samples decreased rapidly
during the first four weeks of implantation although
during this time interval the weight loss remained
negligible.

Copolymers of ε-decalactone and ε-caprolactone
represent useful implant materials only when the decalactone content is very low. A change of the decalactone
content from 8 to 13 mole-% has a pronounced effect on
the biodegradation behavior of implants as shown by the
data of Table II and Fig. 10.

Viscosity changes of both copolymers follow equ.
(6) but observable weight losses occur at an early
stage. The copolymer with the higher decalactone content degraded rather irregularly, very likely because
the implanted tubes had competely lost their shape
already after four weeks in vivo. Excised tubes of the
copolymer with 8 mole-% decalactone content showed only
a slight loss of strength even after 20 weeks.

TABLE II. Biodegradation of Tritium Labeled Poly(Caprolactone-co-ε-decalactone) Capsules[a] in Rabbit

Batch C*2

Time (Weeks)	Intrinsic Viscosity	% Weight Change	% Tritium Change
0	3.21	-	-
4	2.74	0.4	+0.2
	2.76	-0.7	+1.2
8	2.47	+1.0	-0.6
	2.46	-0.7	-0.4
12	2.25	-0.8	-10.1
	2.30	-1.3	-6.8
20	1.95	-1.8	-7.1
	1.92	-2.9	-8.1
28	1.54	-2.8	-15.0
	1.54	-4.9	-9.8

Batch C*3

Time (Weeks)	Intrinsic Viscosity	% Weight Change[c]	% Tritium Change
0	2.09	-	-
4	1.65	0.0	-3.7
	1.75	-1.9	-0.6
8	1.42[b]	-3.8	-18.3
	1.24[b]	-81.1	-81.4
	1.43	-64.0	-76.5
12	1.25	-5.0	-14.7
	1.15	-81.4	-82.9
	1.73	-37.2	-16.8
	1.61	-26.4	-12.3
16	1.45	-38.0	-13.2
	1.41	-23.3	-15.9
20	0.934	-43.0	-49.9
	0.564[b]	-80.8	-84.4
28	0.695	-18.2	-18.9
	0.753	-29.3	-36.8
	0.741	-22.3	-36.4

[a] Dimensions of capsules: length, 2 cm; id, 1.25 mm; od, 1.75-1.90 mm; only caprolactone component is tritium labeled.
[b] Value probably inaccurate because of insufficient amount of polymer for determination.
[c] Capsules of Batch #C*3 recovered as solid, shapeless mass at 4 weeks and later removals.

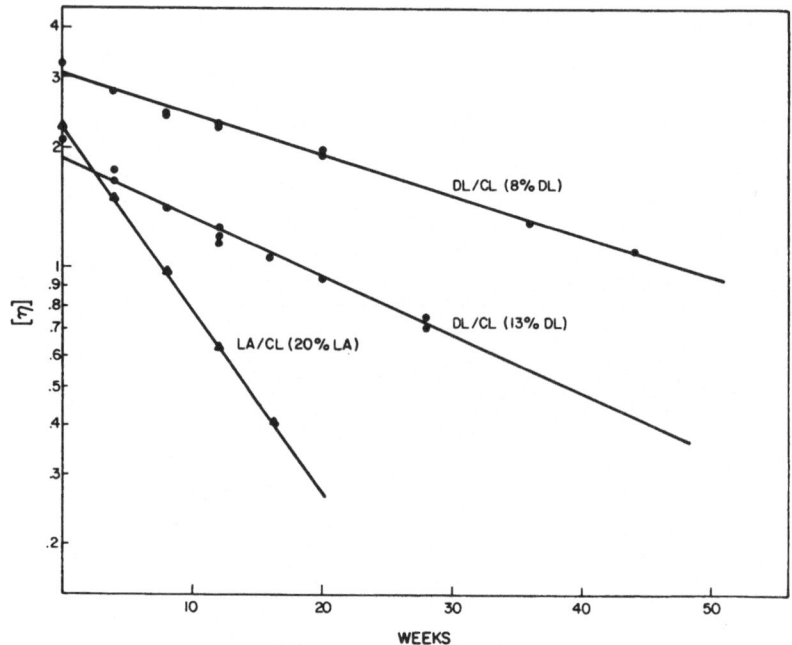

Fig. 10 - In vivo degradation of ε-decalactone-ε-
          caprolactone (DL/CL) and dilactide-ε-
          caprolactone (LA/CL) copolymers in tube
          form. Change in intrinsic viscosity of
          excised implants.

    Viscosity changes during the biodegradation of the
investigated homo and copolymers can be formally described
by equ. (6) with observable weight losses occurring only
after a limiting viscosity is approached. Recognizing
some arbitrariness in defining this limiting viscosity
value (e.g., weight loss $\geq$ 5%), equ. (6) may be used to
estimate the time required for polymers of different
molecular weights to attain the critical value (Fig.
11). In effect, the choice of the initial molecular
weight determines the useful lifetime of the polymer.
It might be pointed out that these estimates will be
valid only for relatively low surface to volume ratios
for which surface erosion plays a minor role. Polymers
in finely dispersed form or in form of extremely thin
films (microcapsules) will be consumed at considerably
higher rates.

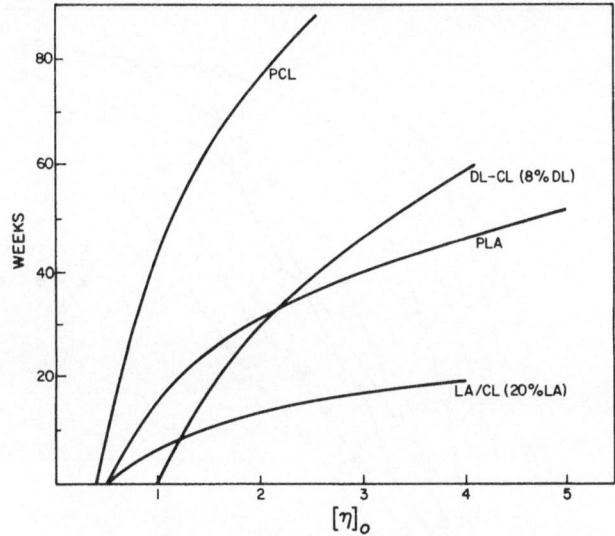

Fig. 11 – Estimated onset of weight loss during in
vivo degradation of homo and copolymers in
dependence on initial intrinsic viscosity.
PCL = poly(caprolactone), PLA = racemic poly-
(lactide), DL/CL = ε-decalactone-co-ε-
caprolactone, LA/CL = racemic dilactide-co-
caprolactone

STEROID RELEASE

Monolithic Devices

Only a limited number of release studies were
undertaken with monolithic devices. In vitro release of
progesterone from poly(caprolactone) films is shown in
Fig. 12 for different loads of steroid (10-30% w/w).
Apart from an initial induction period, the amounts
released are proportional to the square root of time
until about 80% of the steroid are released. In contrast, less than 1% of progesterone was released from
films of poly(DL-lactide) in a comparable time period
(36 hrs.)

In vitro release of progesterone from copolymers of
DL-dilactide and glycolide are summarized in Fig. 13.
Compositions and intrinsic viscosities in benzene were
as follows:

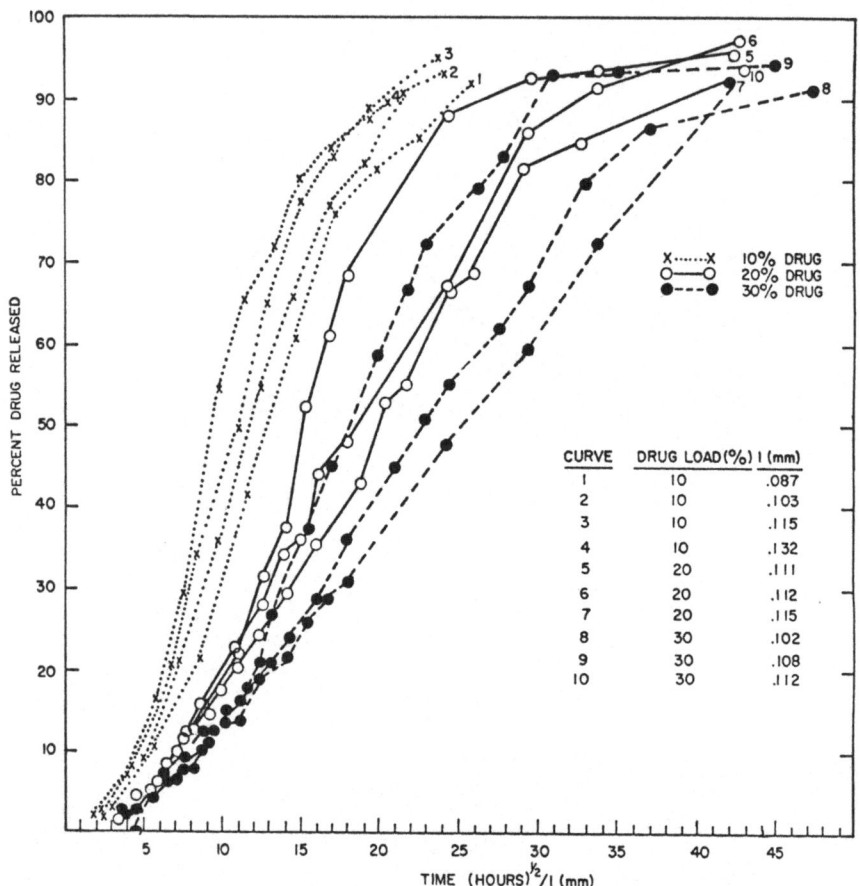

Fig. 12 - In vitro release of progesterone from poly-
(caprolactone) films containing 10-30%
(w/w) drug load

Copolymer	Mole-% Glycolide	Intrinsic Viscosity
C-7	7.0	1.36
C-8	14.0	0.88
C-9	19.4	0.67
C-10	20.8	0.56

None of the films showed significant release for a
period of up to 20 days. At this time the polymer with

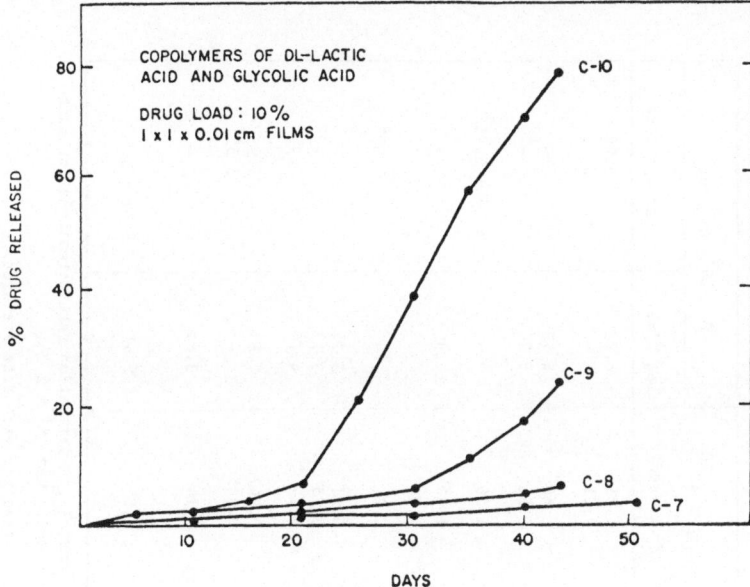

Fig. 13 - In vitro release of progesterone from poly-
(glycolide-co-dilactide) films containing
10% (w/w) steroid. For copolymer identification see text.

the highest glycolide content and possessing the lowest
molecular weight (C-10) began to release progesterone at
a constant rate until it was virtually depleted after 45
days. Copolymer C-9 of the same composition but of
slightly higher molecular weight showed the same behavior
although a significant steroid release started at a
later point (30 days). With both copolymers onset of
drug release coincided with loss in mechanical strength
and fragmentation. Both copolymers C-7 and C-8 remained
intact for the duration of the experiment and drug
release remained minimal although C-8 indicates onset of
increased release after 40 days. In vivo biodegradation
studies of the same copolymers paralleled the in vitro
observations. These results are in contrast to the
previous ones and demonstrate that the release of progesterone from copolymers of dilactide and glycolide is controlled by the degradation of the polymer matrix.

TABLE III. Diffusion and Permeation Coefficients of Steroids in Poly(Caprolactone)[a]

Drug	Vehicle	Film Thickness (cm)	C_s (mg/g)	$J_{lim} \times 10^{10}$ (g/cm·sec)	$D \times 10^9$, cm^2/sec Lag Time	$D \times 10^9$, cm^2/sec Steady State
Testosterone	water	0.0363	4.33	0.35	6.78	8.08
		0.0340		0.31	–	6.73
	sesame oil	0.0139	–	0.66	4.79	–
		0.0363		0.52	5.26	–
Progesterone	water	0.0157	16.6	1.06	–	6.36 ± 0.35[b]
	sesame oil	0.0380	–	1.04	4.96	–
Norethindrone	water	0.0160	2.7[c]	0.21	–	7.68[c]
	sesame oil	0.0120	–	0.17	3.74	–
		0.0125		0.15	3.77	–
		0.0340		0.14	4.74	–
Norgestrel	water	0.0125	1.28	0.047	2.83	3.93
		0.0200		0.045	–	3.55
	sesame oil	0.0115	–	0.14	3.64	–
Ethynylestradiol	water	0.0200	46.6	1.50	2.51	3.32 ± 0.35

[a] C_s = solubility of steroid in polymer; D = Diffusion coefficient; J_{lim} = permeation rate normalized for unit area and unit thickness; all measurements at 37°C; polymer viscosity was $[\eta]$ = 2.24.

[b] These measurements were made over a 9 to 10 day period and the rates for each date averaged.

[c] The C_s for the PCL with $[\eta]$ = 3.17 was assumed for the calculation of D.

Polymer Capsules

Five steroids, Norgestrel, Norethindrone, 17α-ethynylestradiol, testosterone, and progesterone, were studied, with particular emphasis on the latter two. Diffusivities (D), solubilities (c_s), and normalized fluxes ($J_{lim} = Dc_s$) of these steroids for poly(caprolactone) are summarized in Table III. The values were obtained from diffusion cell measurements and equilibrations between polymer films and aqueous steroid solutions.

With all aqueous systems (water and steroid/polymer/-water) the diffusion coefficients were calculated either from lag times (29) or from the steady-state permeation rate and the known saturation concentration of steroid in the polymer. For oil-water systems (sesame oil and steroid/polymer/water) the reported normalized flux is the experimentally observed value and the diffusion coefficient is derived from the lag time.

The reservoir devices consisted of small tubes of 1-2 cm length with internal diameter of 1.2 mm, the outer diameter varying between 1.8 and 3.5 mm. Tritium or carbon-14 labeled steroids were introduced in the form of micronized powders (<5 μm). In most experiments the steroid was dispersed in an inert vehicle before introduction, in order to promote contact between the steroid and the inner capsule wall. The filled tubes were closed by heat sealing.

The amount of steroid released into a large volume of water (ca. 400 ml) at 37°C was measured daily over periods of time up to 150 days. The water was changed daily in order to minimize the steroid concentration which never exceeded 5% of its aqueous solubility and was more typical less then 0.5%. For agitation an Eberbach rotating shaker was used. No change in drug release rate was observed by varying the agitation rate between 135 and 194 rpm.

Constant daily release rates have been achieved in a number of experiments, but more typically the release kinetics have been characterized by an initial rapid decline in release rate during the first 20 days, followed by a more gradual decline during the next 20-150 days. Typical release curves are shown in Fig. 14-

16 which represent the daily release in µg per cm tube
length.

The initial, but not the subsequent decline in
release rate can be largely eliminated by increasing the
wall thickness of the tubes. This is illustrated in
Fig. 17 and 18 for both testosterone and progesterone
release from poly(caprolactone) capsules.

Considerable effort has been expended in identify-
ing and eliminating the origin of rate decline, especially of the pronounced initial one. Diffusion coefficients and permeation rates of progesterone and testosterone through poly(cparolactone) membranes (thickness
100 µm) were found to remain constant when measured in a
diffusion cell over a ten day period. Using a variety
of dispersing agents had no effect on the initial rate
decline. Although the decline was less pronounced in
the absence of dispersing agents, the overall rate was
generally depressed and often erratic under these conditions. The initial decline was also observed if the
time interval between packing and immersion of the
capsules was minimized.

Taken as a whole, these observations strongly
suggest that the release rates of various steroids from
these reservoir devices are determined in part by the
rate of dissolution of the steroid in the dispersing
agent and/or the inner capsule wall. The initial rate
decay may then be related to the dissipation of the
smaller drug particles and already dissolved drug, and
the failure to maintain a constant concentration gradient.
Increasing the wall thickness of the capsules would serve
to eliminate this effect by making the diffusion across
the capsule wall the rate limiting process.

Comparisons of in vitro and in vivo release data
for radiolabeled testosterone from poly(caprolactone)
capsules are presented in Figs. 19 and 20. Prior to
implantation the capsules were conditioned in water for
19 days until the initial decline in release rate had
leveled off. Amounts of drug released in vivo were
determined by summing the radioactivity in urine and
feces (3 or 4 day pools). After about two months the
capsules were excised and drug release under in vitro
conditions was followed for an additional month.

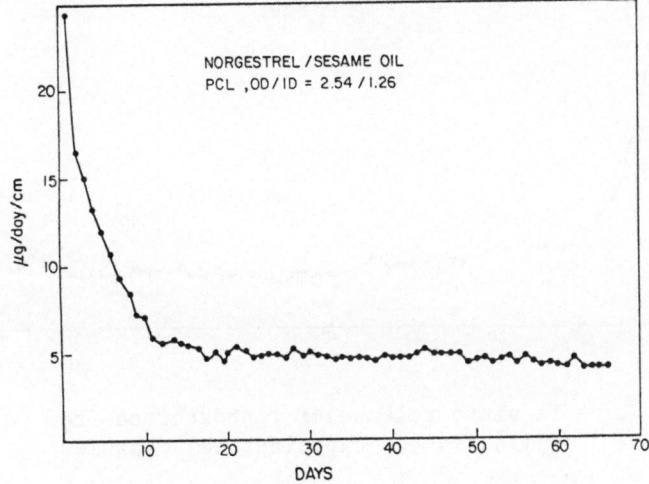

Fig. 14 - In vitro release of Norgestrel from poly-
(caprolactone) capsules

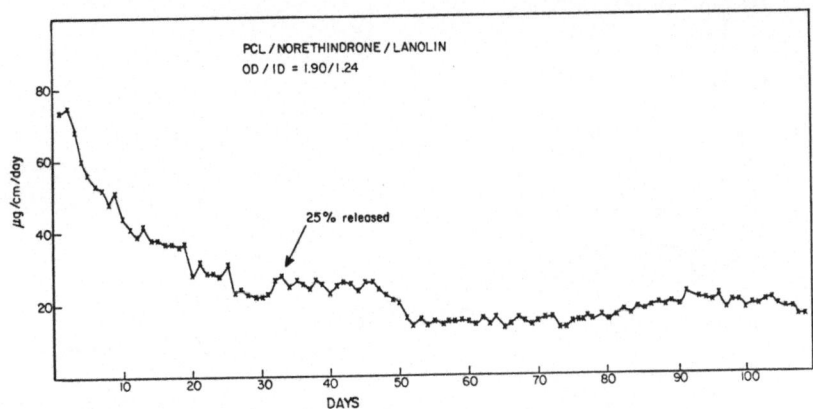

Fig. 15 - In vitro release of Norethindrone from
poly(caprolactone) capsules

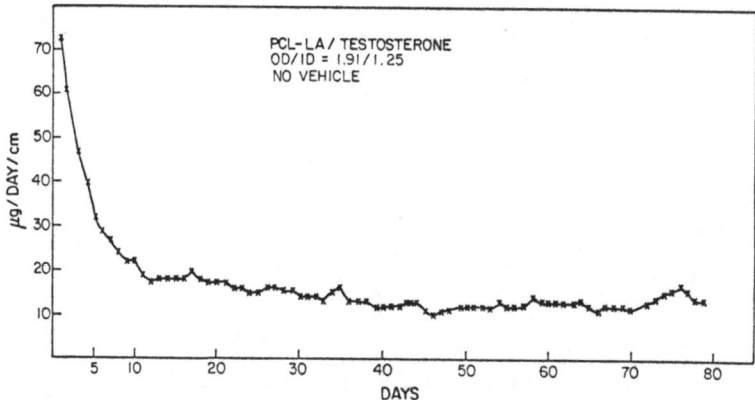

Fig. 16 - In vitro release of testosterone from poly-
(dilactide-co-caprolactone) capsules

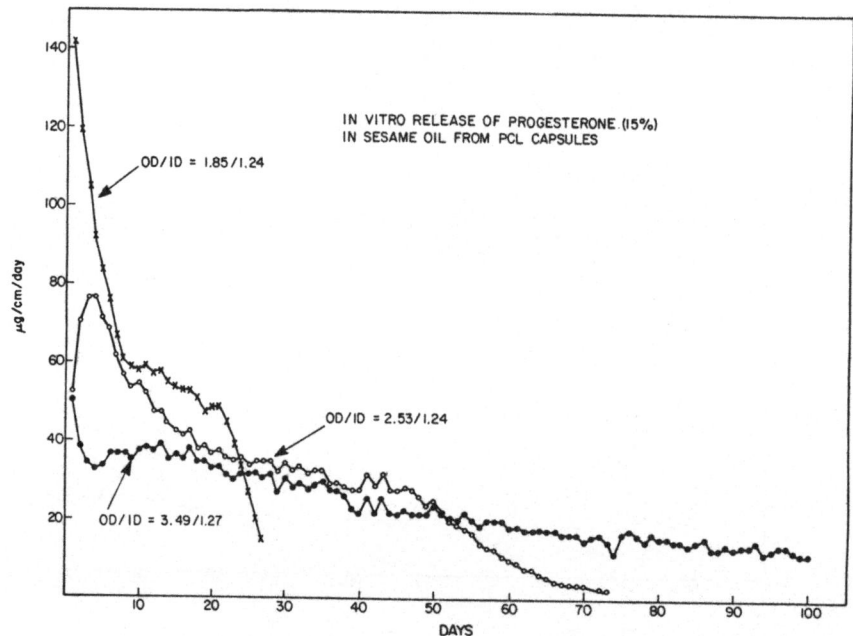

Fig. 17 - Effect of wall thickness on the release rate
of progesterone from poly(caprolactone)
capsules

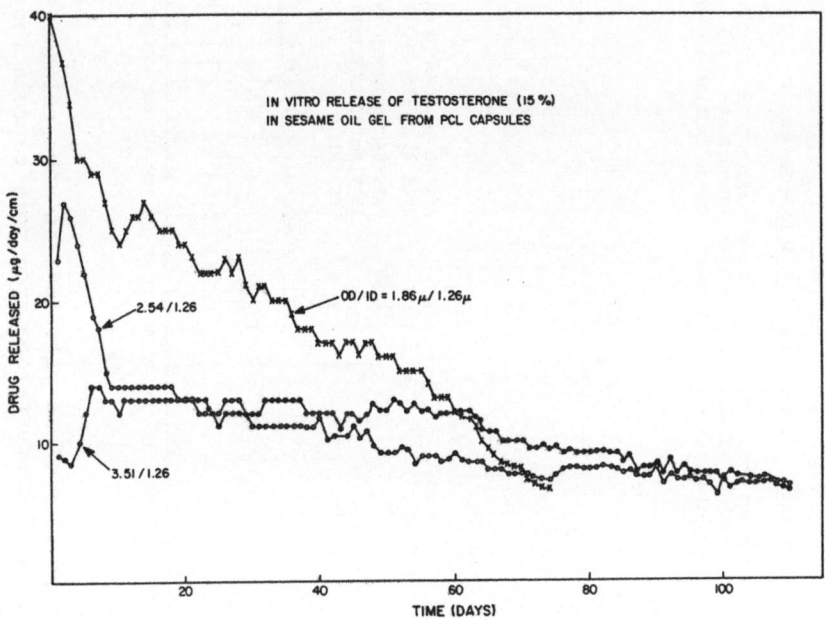

Fig. 18 – Effect of wall thickness on the release rate of testosterone from poly(caprolactone) capsules

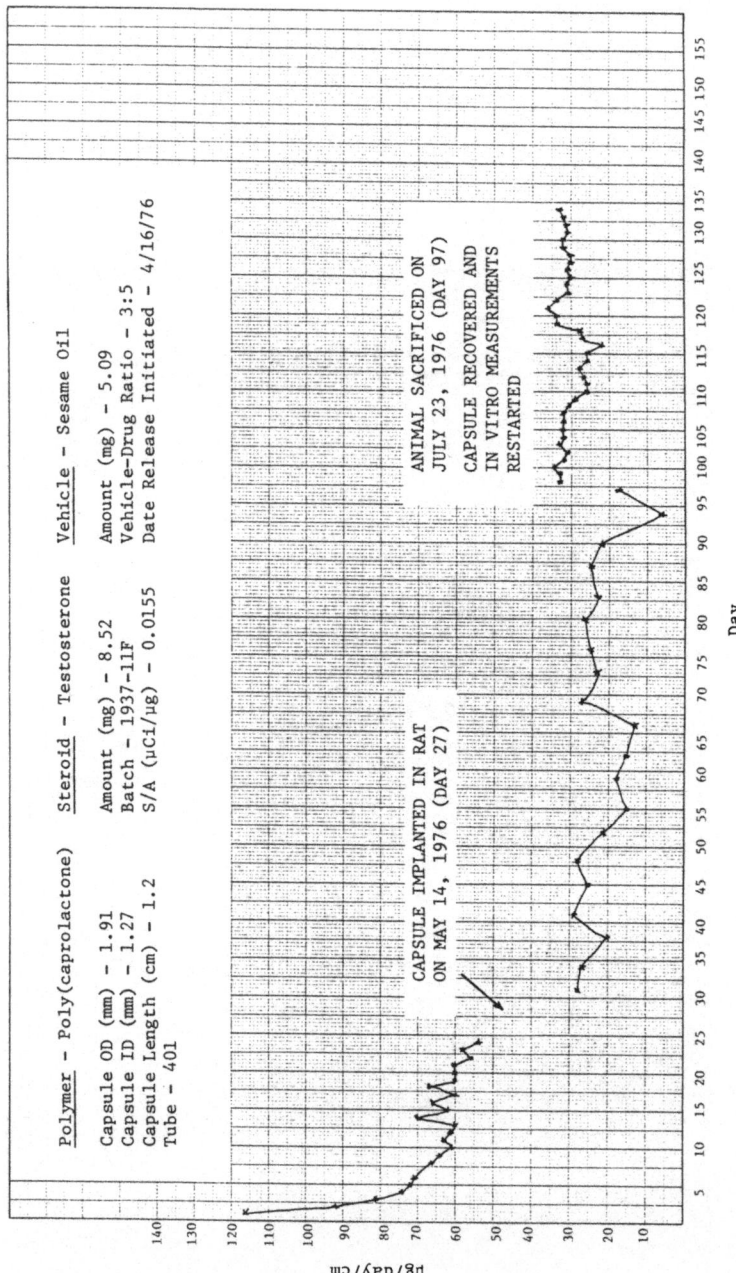

Fig. 19 – Comparison of in vitro and in vivo release
          rates of testosterone from poly(caprolactone)
          capsules

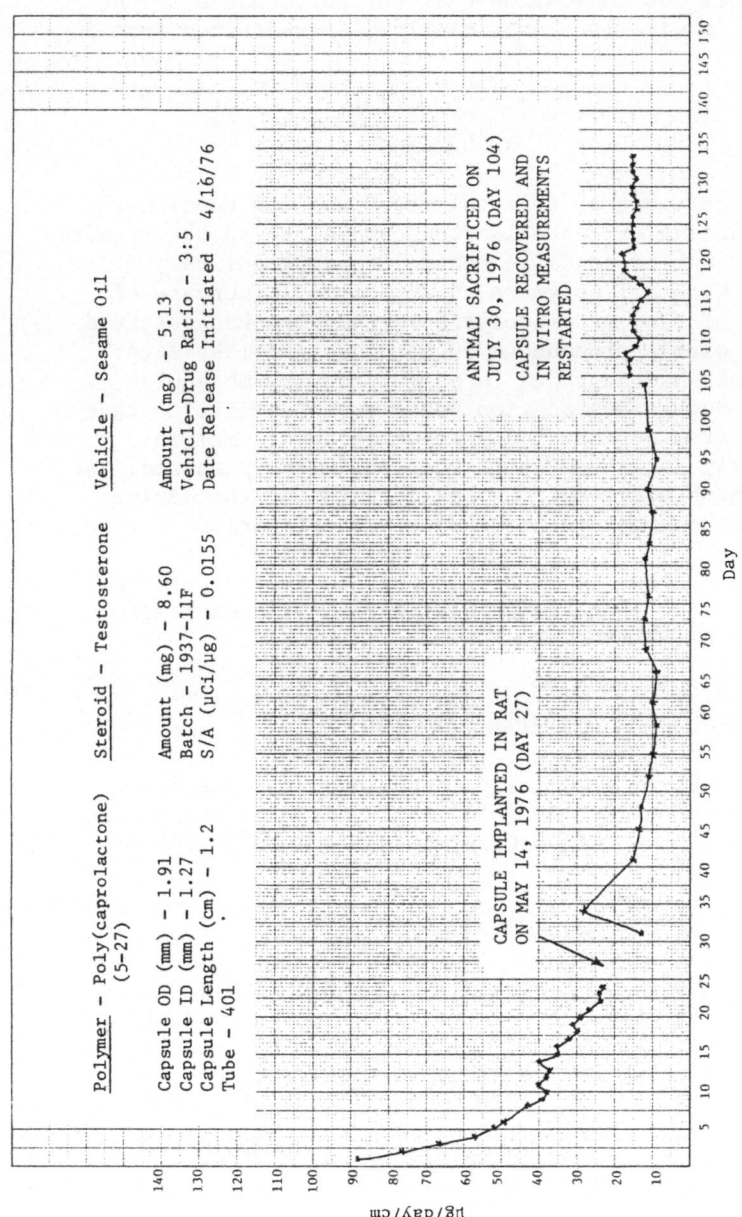

Fig. 20 – Comparison of in vitro and in vivo release rates of testosterone from poly(caprolactone) capsules

Although the in vivo data are more scattered there
is presently no indication that the performance of the
capsules in vivo is significantly different from that in
vitro.

CONCLUSIONS

Homopolymers of ε-caprolactone and its copolymers
with dilactide or ε-decalactone come close to the requirements for a biodegradable controlled drug delivery
system. With all polymers an exponential increase of
the rate of random, hydrolytic degradation was observed
together with a pronounced onset of weight loss after
the polymer had degraded to a limiting low molecular
weight. Such a degradation behavior is similar to that
required from an ideal polymer which should remain
practically unchanged until the device is exhausted. At
this point biodegradation should start and the device
should be consumed in a relatively short period of time.

With changes in molecular weight, and composition
in the case of copolymers, a wide span of useful lifetimes can be covered.

ACKNOWLEDGMENT

The authors gratefully acknowledge the support of
this work by the National Institute of Child Health and
Human Development under Contract No. N01-ND-3-2741.

REFERENCES

1. T. Higuchi, J. Pharm. Sci., $\underline{50}$, 874 (1961).
2. R. W. Baker and H. K. Lonsdale, "Controlled Release:
    Mechanism and Rate", in "Controlled Release of
    Biologically Active Agents", A. C. Tanquary and
    R. E. Lacey, Eds., Plenum Press, New York, 1974.
3. J. Folkman and D. M. Long, J. Surg. Res., $\underline{4}$, 139 (1964).
4. Compare Ref. 8-30 in Ref. 2.
5. J. H. R. Woodland, S. Yolles, D. A. Blake, M. Helrich,
    and F. J. Meyer, J. Med. Chem., $\underline{16}$, 897 (1973).

6. S. Yolles, J. Eldridge, T. D. Leafe, J. H. R. Woodland, D. R. Blake, and F. J. Meyer, "Long Acting Delivery Systems for Narcotic Antagonists", in "Controlled Release for Biologically Active Agents", A. C. Tanquary and R. E. Lacey, Eds., Plenum Press, New York, 1974.

7. T. D. Leafe, S. F. Sarner, J. H. R. Woodland, S. Yolles, D. A. Blake, and F. J. Meyer, Adv. Biochem. Psychopharm., 8, 569 (1974).

8. S. Yolles, T. D. Leafe, J. H. R. Woodland, and F. J. Meyer, J. Pharm. Sci., 64, 384 (1975).

9. S. Yolles, T. D. Leafe, and F. J. Meyer, J. Pharm. Sci., 64, 115 (1975).

10. S. Yolles, "Controlled Release of Biologically Active Agents" in "Polymers in Medicine and Surgery", R. L. Kronenthal, S. Oser, and E. Martin, Eds., Plenum Press, New York.

11. T. M. Jackanicz, H. A. Nash, D. L. Wise, and J. B. Gregory, Contraception, 8, 227 (1973).

12. C. Thies, "Development of Injectable Microcapsules for Use in the Treatment of Narcotic Addiction", in Natl. Inst. on Drug Abuse Res. Monograph No. 4, R. Willette, Ed., 1976.

13. N. Mason, C. Thies, and T. J. Cicero, J. Pharm. Sci., in press.

14. J. Kleine and H. H. Kleine, Makromol. Chem., 30, 23 (1959).

15. W. Dittrich and R. C. Schulz, Angew. Makromol. Chem., 15, 109 (1971).

16. G. L. Brode and J. V. Koleske, "Lactone Polymerization and Polymer Properties", in "Polymerization of Heterocyclics", O. Vogl and J. Furukawa, Eds., Marcel Dekker, New York, 1973.

17. H. Cherdron, H. Ohse, and F. Korte, Makromol. Chem., 56, 179 (1962).

18. T. Ouhadi, Ch. Stevens, and P. Teyssie, Makromol. Chem., Suppl. 1, 191 (1975).

19. C. H. Holten, A. Muller, and D. Rehbinder, "Lactic Acid", Verlag Chemie, Weinheim, 1971.

20. A. Schindler and D. Harper, Polymer Letters, 14, in press.

21. S. R. Palit and B. M. Mandal, J. Macromol. Sci., C2, 225 (1968).

22. R. D. Lundberg, J. V. Koleske, and K. B. Wischmann, J. Polymer Sci., A-1, 7, 2915 (1969).

23. S. Iwabuchi, V. Jaacks, and W. Kern, Makromol. Chem., 177, 2675 (1976)

24. O. Ringer and A. Skrabal, Mh. Chem., 43, 507 (1922).

25. A. Charlesby and S. H. Pinner, Proc. Royal Soc., (London), A249, 376 (1959).

26. T. J. Roseman, J. Pharm. Sci., 61, 46 (1972).

27. T. G. Fox, Bull. Am. Phys. Sco., 2, 123 (1956).

28. D. A. S. Ravens and I. M. Ward, Trans. Faraday Soc., 57, 150 (1961).

29. R. M. Barrer, "Diffusion In and Through Solids", Cambridge, University Press, 1941.

DISCUSSION

H. MORAWETZ - POLYTECHNIC INSTITUTE OF NEW YORK.
What about the unnatural isomers of lactic acid?
A. SCHINDLER - RESEARCH TRIANGLE INSTITUTE. It was shown
by Brady et al. [J. Biomed. Mater. Res., 7, 155 (1973)]
that DL-polylactic acid as an implant is well tolerated by
the host tissue and is most likely eliminated as respired
carbon dioxide. One has also to consider that homopolymers
of racemic dilactide degrade at such a low rate that the
degradation products can be easily accommodated by the
surrounding tissue. Extremely fast degrading copolymers
might yield adverse tissue reactions due to the high local
acid concentration.

H. MORAWETZ. You talked about atactic polymer. Your
atactic polymer was made from a mixture of l,l and d,d
lactides. Is it possible to start with a d,l lactide?
A. SCHINDLER. There exists also a meso form of dilactide
(m.p. = 42°C) which is not commercially available.
Polymerization without inversion would yield a polymer
with a stereosequence distribution identical to that of
racemic poly(lactide) but with m and r dyads being mutually
exchanged, i.e., (mmm) would become (rrr), (mrm) would
become (rmr), etc. The polymer would not be truely atactic
since m dyads have to be separated by an odd number of r
dyads. In the polymerization of dilactide initiated with
amines one can assume ring opening and propagation to occur
via a carboxylate ion as in the case of pivalolactone.
Inversion of one asymmetric carbon during the propagation
step is then possible, and, consequently, a greater
randomization of m and r dyads will occur.

H. MORAWETZ. Did you make a study with that material?
A. SCHINDLER. We prepared several poly(lactides) from
dilactide by initiation with triethylamine. The polymers
derived from L-dilactide were practically optically inactive
with specific rotations of about -10° versus -162° for
poly(L-lactide). All polymers had the same high T_g value
as those obtained from racemic dilactide with stannous
octoate. Apparently, further randomization of m and r
dyads has no effect on T_g.

N. DODDI - ETHICON, INC. You have made copolymers
from lactic acid and caprolactone and you said that you
have not observed any tissue reactions. Have you looked
at any long range effects of these copolymers, e.g., have
you looked at the carcinogenicity of these polymers?
A. SCHINDLER. Caprolactone could be derived from the
implants only if present as residual monomer not removed
during polymer isolation and purification. The product of
degradation will be hydroxyhexanoic acid which will not
form lactone in an aqueous system. A great number of higher
lactones are also used as food additives in 10-100 ppm
quantities, and therefore, cannot be considered as
carcinogenic.

N. DODDI. You were talking about poly-d,l-lactide.
You said that the polymer may be a soft material. We made
this polymer and found it to be a hard glassy polymer.
A. SCHINDLER. Yes, it is. We hoped, however, that by
decreasing the average length of all-m sequences in racemic
poly(lactide), i.e., by preparing a truely atactic polymer,
the T_g value could be decreased. Unfortunately, this was
not observed.

R. CAPOZZA - ALZA RESEARCH. One of the drug release
profiles you showed displayed release of drug from a
polycaprolactone rod; I believe you had an induction period
followed by an accelerated rate of release. I think what
is happening in that situation is that drug was diffusing
from the rod followed by an accelerated rate of erosion of
the system. Thus, there is little or no erosion in the
early stages of drug release followed by an accelerated
rate of erosion and release of drug.   A. SCHINDLER. You
are referring to monolithic devices prepared from dilactide-
glycolide copolymers shown in Fig. 13. These devices show
typical behavior for erosion controlled drug release.
Appreciable drug release coincided with onset of erosion.

R. CAPOZZA. Yes, maybe it was one of the copolymers.
I think you will find that hydrolysis of an ester follows
autocatalytic hydrolysis.   A. SCHINDLER. I completely
agree with this statement and I believe its validity was
amply demonstrated by all our in vitro and in vivo

degradation studies. Indeed, it is the autocatalytic nature
of ester hydrolysis which permits the use of polyesters for
biodegradable reservoir devices. The ideal material for
such devices should remain unchanged until the device is
exhausted of drug when rapid degradation should commence.
An autoaccelerated degradation kinetics seems to be the best
approach toward these requirements provided the rate constant
can be modified according to needs by modifications of the
polymer. This has been demonstrated by data shown in
Fig. 11.

DESIGN OF TRANSDERMAL THERAPEUTIC SYSTEMS

S.K. Chandrasekaran, J.E. Shaw

ALZA Corporation

950 Page Mill Rd., Palo Alto, CA 94304

INTRODUCTION

For the vast majority of therapeutic agents, the efficacy of drug action and minimization of unwanted pharmacological effects are the therapist's two goals. Accordingly,
the rational aim of pharmaceutical development should be
to produce dosage forms whose release kinetics are defined
by the requirements of meeting the therapist's goals. Thus,
in the development of a controlled drug delivery system,
while the functional lifetime of the system is important,
the temporal pattern of drug release rate becomes the premier
criterion of design.

Therapeutic systems are drug dosage forms whose metrics
are rate and duration of drug delivery. In the most
general sense, a therapeutic system consists of a drug component, a drug delivery module, a platform, and a therapeutic program. The drug component is selected on the
basis of its established therapeutic effectiveness and appropriate pharmacokinetic behavior. The drug delivery
module is subdivided into four elements, namely: a drug
reservoir, a rate controller, an energy source, and a
delivery portal. The platform houses the drug and delivery
module and couples the therapeutic system to the selected
body site. The therapeutic program, which is executed by
the drug delivery module, presents the drug in the most
beneficial and reliable manner (1-4).

By using the intact skin as a drug entry portal,
transdermal therapeutic systems can deliver certain drugs,

chosen for their safety and specificity, to the
systemic circulation in a more convenient and effective
way than is currently available. The transdermal route
offers several potential advantages for the systemic delivery of drugs:

    1. Elimination of the vagaries of gastrointestinal
absorption normally affecting drugs taken orally;

    2. Reduction of drug metabolism due to an initial
bypass of the liver;

    3. Controlled administration of drugs with small
therapeutic indices;

    4. Elimination of the pulse entry of drugs to the
systemic circulation, thereby reducing side effects;

    5. Utilization of drugs with short half-lives that
otherwise could not be successfully delivered to maintain
therapy.

An understanding of the factors which influence cuta-
neous and percutaneous absorption, together with advances
in polymer technology, have led to the possibility of providing optimal drug therapy for the treatment of systemic
disorders -- acute or chronic -- by means of controlled,
unattended administration of a drug to the skin surface,
for prolonged, predetermined periods of time (5).

GENERAL PERMEATION PROPERTIES OF SKIN

The skin is one of the most extensive and readily
accessible organs of the human body; only a fraction of a
millimeter of tissue separates the skin surface from the
underlying capillary network. The principal resistance to
penetration of drugs and other small molecules through
intact human skin resides within the stratum corneum, which
is comprised of dead, keratinized, partially desiccated
epidermal cells. The stratum corneum is a heterogeneous
structure containing about 40% protein, about 40% water and
about 20% lipid in a layered, closely packed array of flat-
tened, interdigitated cells (6).

The permeation of drugs and other micromolecules through
intact human skin occurs principally by diffusion as described by Fick's Law, with the gradient in drug concentration
across the entire skin being localized within the stratum
corneum (6). The inherent low permeability of the skin
varies in degree, depending on the nature of the agent
topically applied and the treatment of the skin prior to

such application. Factors which affect permeation of
drugs through skin include: the physicochemical properties of the therapeutic agent, the vehicle of the
formulation, and occlusion and hydration of the stratum
corneum (6).

   We have developed an _in vitro_ experimental procedure to
measure the permeation characteristics of drugs through skin
(7). The first drug to pass through the development activity is scopolamine, which is a belladonna alkaloid with a
pK of 7.35. For scopolamine, skin permeability is strongly
pH dependent, in that the nonionic (more lipophilic) form

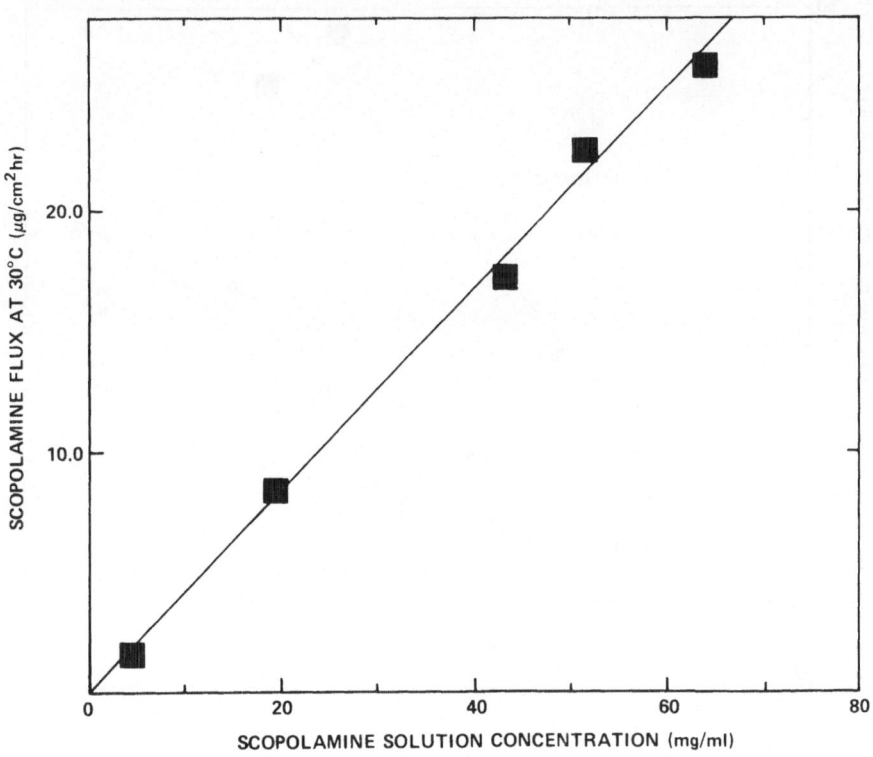

Figure 1. Effect of Concentration on Scopolamine Flux
Through Human Epidermis (Epidermis A)

of the drug is decidedly more skin permeable than the
ionic form.

The transdermal steady state flux of scopolamine is a
linear function of the concentration of the aqueous drug
solution contacting the stratum corneum surface of the skin
(Figure 1). An equilibrium sorption isotherm for scopolamine
in the same tissue is shown in Figure 2. The steady state
diffusivity, determined by dividing the measured steady
state flux by the computed gradient in the stratum
corneum of dissolved drug, is essentially independent of
drug concentration in the solution contacting the stratum

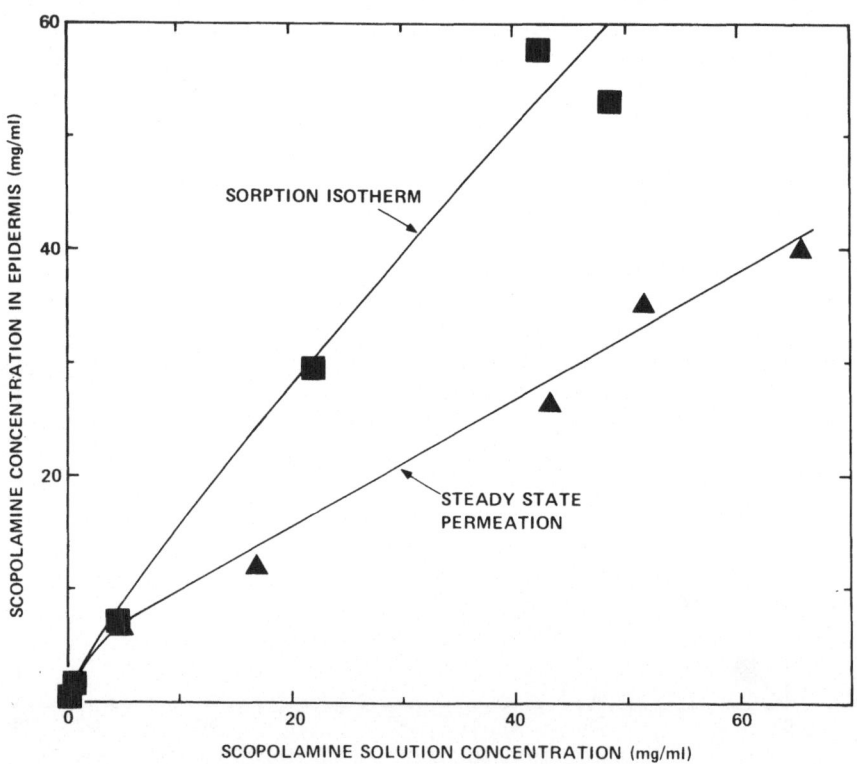

Figure 2. Variation of Drug Concentration in Epidermis
with Aqueous Solution Concentration (Epidermis A)

TABLE 1

Scopolamine Diffusion Coefficients (Epidermis A)

SCOPOLAMINE SOLUTION CONCENTRATION C, mg/ml	STEADY STATE DIFFUSION COEFFICIENT, D_{SS} $cm^2/sec \times 10^{10}$
64.0	5.0
51.4	5.2
43.1	4.8
19.5	5.0
4.4	4.6

corneum (Table 1). A simplistic model of the sorption process, which invokes the coexistence of dissolved and mobile sorbed molecules in equilibrium with site bound and immobile molecules within the tissue, quite accurately correlates experimental sorption data and transient transport measurements (8).

DEVELOPMENT OF A TRANSDERMAL THERAPEUTIC SYSTEM

Since the permeability of skin to scopolamine is sufficiently high to permit transdermal administration at therapeutically useful rates, it has been possible to design a membrane modulated delivery system, which delivers scopolamine at a predictable and constant rate.

The Transdermal Therapeutic System-scopolamine, shown schematically in Figure 3, is a multilayer laminate. It is comprised of a steady-state drug reservoir containing scopolamine in a polymeric gel, sandwiched between an impermeable backing membrane and a rate-controlling, microporous membrane. On the dermal side of the rate-controlling mem-

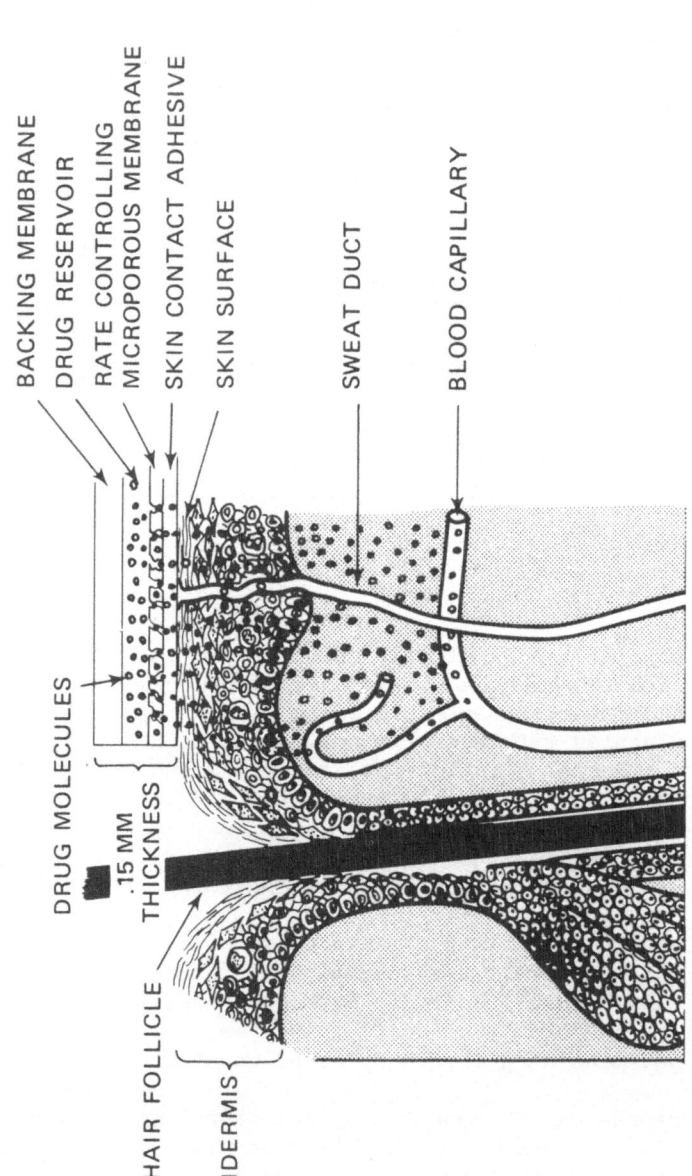

Figure 3. Schematic Diagram of a Transdermal Therapeutic System in Place on Surface of Intact Skin

brane is an adhesive gel containing scopolamine; this gel
layer serves both as an adhesive to secure the system on
the skin surface and as a priming dose drug reservoir to provide an initial priming dose of drug prior to the establishment of a controlled input of drug to the skin surface. The
magnitude of the priming dose was chosen in light of the
degree of drug immobilization in the skin and the pharmacokinetic properties of scopolamine. The protective
strippable film is removed just prior to the application of
the system to the skin surface.

A. Characteristics of In Vitro Delivery of
Scopolamine from the Transdermal Therapeutic
System-scopolamine into an Infinite Sink

Theory. The transport characteristics of scopolamine
from the system are determined by molecular diffusion through
the various elements of the multilayer laminate. During the
priming dose period, drug diffusion from the contact adhesive
layer dominates the temporal pattern of drug release. However, during steady-state delivery, rate-limitation, or control, is resident in the microporous membrane.
Drug Diffusion in the Microporous Membrane. The pores
of the microporous membrane are filled with a suitable solvent having the necessary solubility and diffusion characteristics for scopolamine.

The fundamental equation governing scopolamine transport through the microporous membrane shown schematically
in Figure 4 is: (9)

$$J = \frac{\varepsilon \, D \, \Delta C}{\tau \ell} \qquad (1)$$

Where J = flux in mass/unit area/unit time
         D = diffusion coefficient
         ΔC = concentration decrement across the membrane
         ε = membrane porosity
         τ = membrane tortuosity
         ℓ = membrane thickness

If the drug concentration in the contact adhesive side of
the membrane is small compared to the saturated drug concentration in the drug reservoir side of the membrane, Equation
(1) could be reduced to Equation (2).

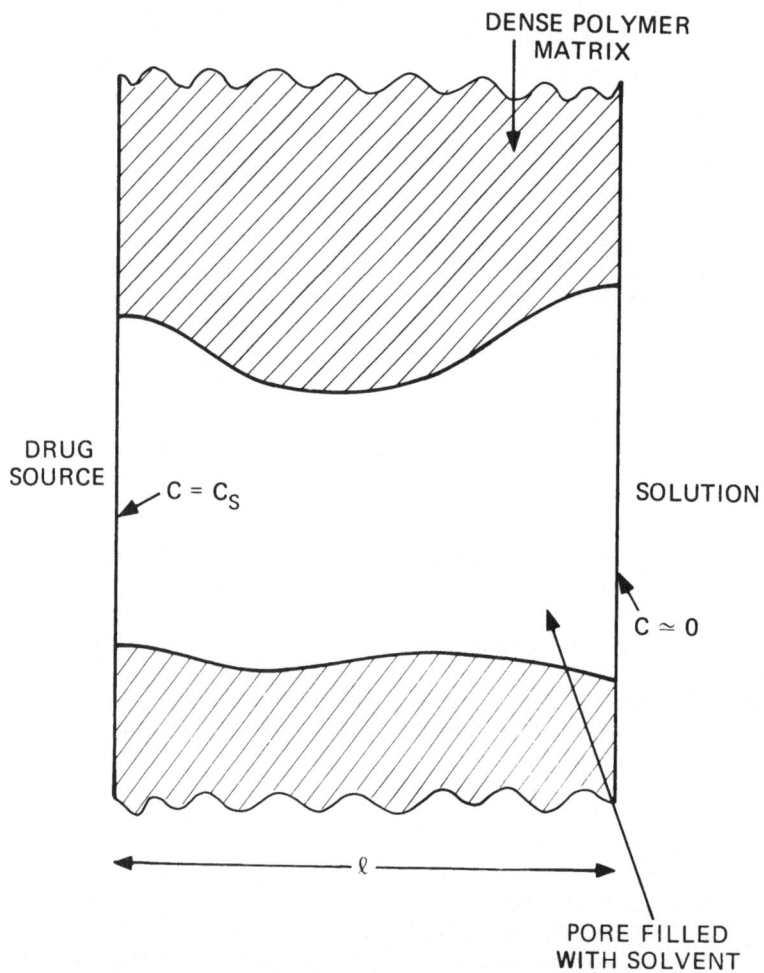

Figure 4. Schematic Representation of Transport Through Microporous Membranes

$$J = \frac{\varepsilon \, D \, C_S}{\tau \ell} \tag{2}$$

where C_S is the scopolamine concentration (solubility) in the drug reservoir.

<u>Drug Diffusion in the Contact Adhesive.</u> The contact adhesive formulation consists of a polymeric gel saturated with scopolamine, with excess drug uniformly dispersed throughout the gel. The release kinetics from such a dispersed formulation has been derived previously by Higuchi (10) using the model shown in Figure 5.

Applying Fick's Law for the slab geometry and neglecting boundary layer resistances in the solution, we have:

$$\frac{d \, M_t}{dt} = \frac{A \, D \, C_S}{x} \tag{3}$$

where M is mass, t is time, A is area, C_S is concentration (solubility), D is diffusion coefficient, and x is incremental distance.

At time t from mass balance considerations,

$$\frac{x}{\ell} = \frac{M_t + Ax \, C_S / 2}{M_\infty} \tag{4}$$

where ℓ is the thickness.

Combining Equations (3) and (4), we obtain,

$$\frac{d \, M_t}{dt} = A \, D \, C_S \, \frac{M_\infty}{M_t} \left[\frac{1}{\ell} - \frac{A \, C_S}{2M_\infty} \right] \tag{5}$$

Integrating Equation (5)

$$M_t^{\,2} = 2 \, A \, D \, C_S \left[\frac{1}{\ell} - \frac{A \, C_S}{2M_\infty} \right] M_\infty \, t \tag{6}$$

however,

$$M_\infty = A \, C_o \, \ell \tag{7}$$

where C_O is the total drug concentration (dissolved plus dispersed).

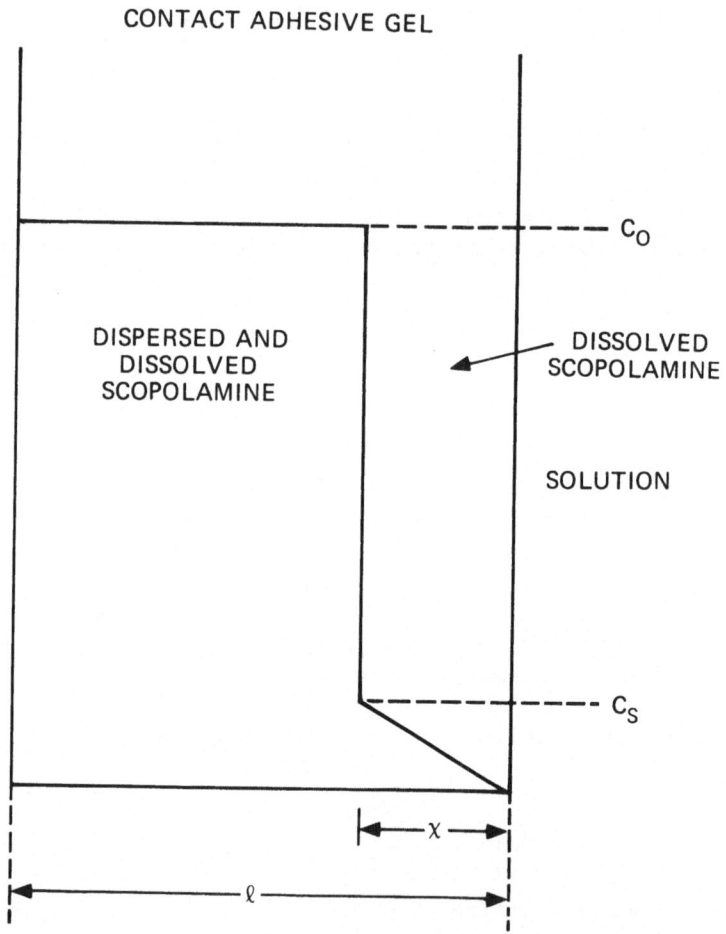

Figure 5.   Idealized Model for Dispersed Drug System

Substituting Equation (7) into Equation (6), we obtain:

$$M_t = A[Dt \, C_S(2C_o - C_S)]^{\frac{1}{2}} \tag{8}$$

For $C_o \gg C_S$, Equation (8) reduces to:

$$M_t = A[2Dt \, C_S \, C_o]^{\frac{1}{2}} \tag{9}$$

The release rate at any time, t, is given by:

$$\frac{1}{A}\frac{d \, M_t}{dt} = \left[\frac{D \, C_S \, C_o}{2t}\right]^{\frac{1}{2}} \tag{10}$$

The point of exhaustion (or the time when the last solid drug just dissolves, t_∞) can be computed as:

$$t_\infty = \frac{\ell^2 \, C_o}{2D \, C_S} \tag{11}$$

Hence, the contribution of the drug in the contact adhesive to the overall scopolamine delivery kinetics varies as $t^{-\frac{1}{2}}$, until the time when the contact adhesive becomes depleted of excess drug.

A typical in vitro release rate-time profile of scopolamine from such a therapeutic system into an infinite sink at isotonic and isothermal conditions is shown in Figure 6. The data points represent values experimentally measured, and the solid line represents the profile predicted by theory, and it is apparent that the agreement between the two is good.

To follow the rate of drug input to the systemic circulation during the in vivo application of the therapeutic system, we monitored the rate of urinary excretion of the drug, using a sensitive and specific assay (11). Taking into account the fact that following intramuscular or intravenous administration of scopolamine, only 10% of the drug is recovered in the urine in the free form, we observe that within about four hours following application of the system, the urinary excretion rate of drug approaches 10% of the release rate of drug from the system in vitro (Figure 7). The provision of the priming dose of drug serves to saturate the immobilization sites for scopolamine within the stratum cor-

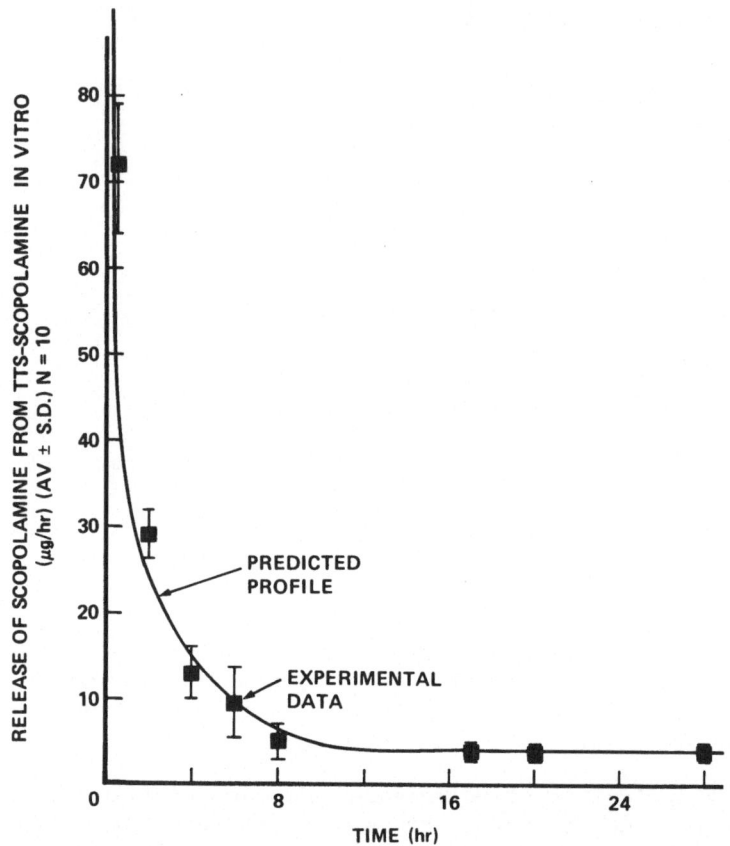

Figure 6.   Scopolamine Release Rate Profile: Comparison of
Theory and Experiment

neum and hence permits rapid establishment of steady state
in urinary excretion rate of drug, confirming that the drug
administration kinetics are indeed governed by the therapeutic system and not by patient related factors.

The present Transdermal Therapeutic System-scopolamine
is 2.5 cm^2 in area, and its strength is specified by its
temporal pattern of drug release: 200 μg priming dose, 10
μg/hr for 72 hours. Results of large scale clinical
studies have indicated that the system is a safe and effec-

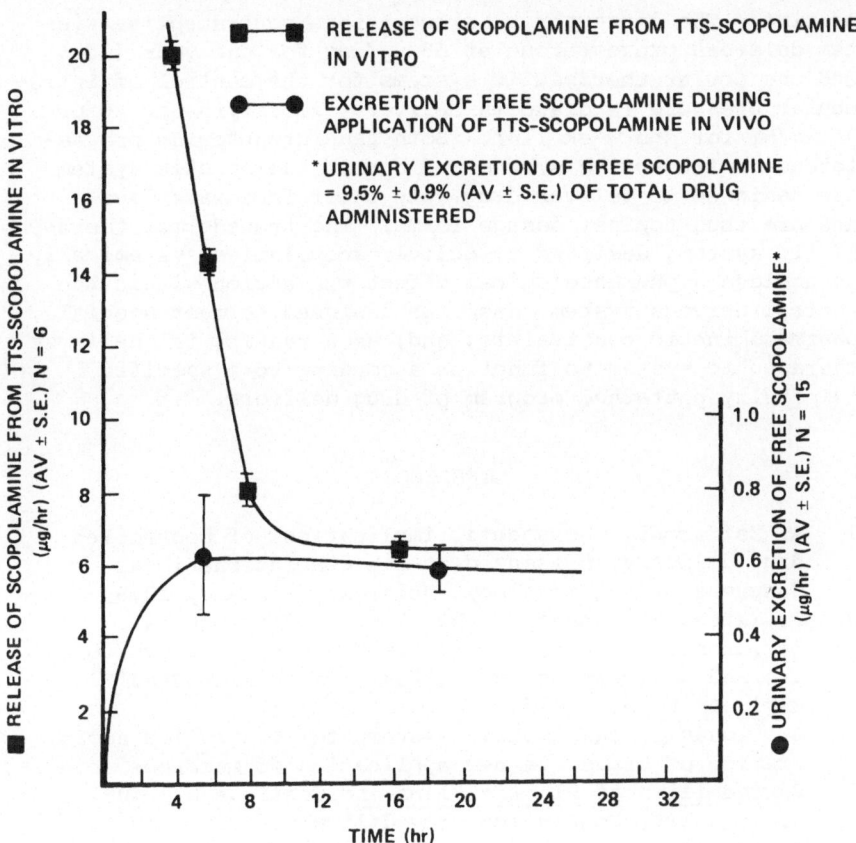

Figure 7. Functionality of TTS-Scopolamine In Vitro and In Vivo

tive dosage form for systemic administration of scopolamine for prevention of motion-induced nausea (12).   The temporal pattern of systemic delivery permits the anti-emetic activity of scopolamine to be realized, with minimal incidence of other parasympatholytic effects of the drug.

CONCLUSIONS

Therapeutic systems have previously been developed to deliver drugs at a constant rate and for an extended period

of time. The intrauterine progesterone contraceptive system delivers progesterone at 65 µg/day for one year (13),
and the ocular therapeutic systems for the control of intra-
ocular pressure in glaucoma delivers pilocarpine at 20 or
40 µg/hr for one week (14). Both the intrauterine progesterone contraceptive system and ocular therapeutic systems
are designed to deliver drug into their immediate locale,
and are thus topical dosage forms. The transdermal therapeutic system, designed to deliver scopolamine systemically
to achieve a pharmacological effect via action within the
central nervous system, has been designed to meet special
pharmacokinetic constraints, and, as a result, is the first
therapeutic system to function according to a specific
temporally patterned program of drug delivery.

REFERENCES

1. A. Zaffaroni, Therapeutic implications of controlled
    drug delivery, presented at the 6th International
    Congress of Pharmacology, Helsinki, Finland, 1975.
2. F. Yates, H. Benson, R. Buckles, J. Urquhart, A.
    Zaffaroni. Designs for improved therapy through con-
    trolled delivery of drugs, Advances in Biomedical
    Engineering 5: 1975.
3. A. Michaels, Therapeutic systems for controlled adminis-
    tration of drugs: a new application of membrane science,
    Permeability of Plastic Films and Coatings 6: 409, 1975.
4. J. Urquhart, Controlled drug delivery systems in the
    treatment of hyperendemic diseases in developing
    countries, IEEE Systems, Man, and Cybernetics Society,
    Proceedings of the 1974 International Conference 197:
    1974.
5. J. Shaw, S. Chandrasekaran, A. Michaels, L. Taskovich.
    Controlled transdermal delivery, In Vitro and In Vivo,
    Animal Models in Dermatology , Churchill Livingstone,
    New York, 1975.
6. R. Scheuplein, I. Blank. Permeability of the skin,
    Physiol. Rev., 51: 702, 1971.
7. A. Michaels, S. Chandrasekaran, J. Shaw. Drug permeation
    through human skin: Theory and In Vitro experimental
    measurement. AIChE J., 21: 985, 1975.
8. S. Chandrasekaran, A. Michaels, P. Campbell, J. Shaw.
    Scopolamine permeation through human skin In Vitro.
    AIChE J., 22: 828, 1976.

9. J. Crank, The Mathematics of diffusion, Oxford University Press, New York, 1970.

10. T. Higuchi, Rate of release of medicaments from ointment bases containing drugs in suspension. J. Pharm. Sci. 50: 874, 1961.

11. W.F. Bayne, F.T. Tao and N. Crisalogo. Submicron assay for scopolamine in plasma and urine. J. Pharm. Sci. 64: 288, 1975.

12. J. Shaw, L. Schmitt. Clinical pharmacology of scopolamine in prevention of motion-induced nausea. (In preparation) 1976.

13. B. Pharriss, R. Erickson, J. Bashaw, S. Hoff, V. Place and A. Zaffaroni. Progestasert: A uterine therapeutic system for long term contraception: I. Philosophy and clinical efficacy, Fertil. Steril. 25: 922, 1974.

14. J. Shell and R. Baker, Diffusional systems for controlled release of drugs to the eye. Ann. Ophthalmol. 6: 1037, 1974.

DISCUSSION

D. J. SAVAGE - EASTMAN KODAK CO. Is the FDA currently
looking at these products? S. K. CHANDRASEKARAN - ALZA
RESEARCH. The New Drug Application has been filed on the
Transdermal Therapeutic System-scopolamine, and based on
our last interaction with the FDA, it appears that everything
is in order and the application is going through the
bureaucratic process.

D. J. SAVAGE. In the beginning of your lecture you said
that a drug gets metabolized during its passage through the
liver. (1) Is this very common, and (2) does one usually
take much more of a particular drug than one really needs?
S. K. CHANDRASEKARAN. (1) Yes, generally so. (2) Yes, this
is true. Just as a general comment, alcohol can interfere
with the enzymatic processes in the liver and it is possible
to get much higher concentrations from the same dose.

D. J. SAVAGE. Can essentially any drug be delivered by
this system? S. K. CHANDRASEKARAN. No. As I had explained
earlier, the molecular weight of the drug, its physicochemical properties, and the dose range in which the drug
is therapeutically effective, are important considerations in
the choice of the drug candidate.

R. MOORE - EASTMAN KODAK CO. Will you comment on the
skin adhesive that was used? S. K. CHANDRASEKARAN. Yes.
It is essentially a nonaqueous formulation containing
polyisobutylene - it serves as an adhesive and at the same
time offers minimal transport resistance to the drug.

R. MOORE. In the case of L-Dopa, there are preparations
on the market now which inhibit enzymatic degradation. I
wonder if you might know whether this material lends itself
either to the transdermal or topical delivery systems?
S. K. CHANDRASEKARAN. We really have not looked at that,
so I cannot comment.

R. C. SUTTON - EASTMAN KODAK CO. I would assume that
since the devices come into intimate contact with the skin,
the environmental effects on diffusion rate would be minimal.
Is that correct? S. K. CHANDRASEKARAN. Yes. The main
reason for choosing the microporous membrane is that the

activation energy for drug transport is generally low, in the vicinity of 5 kcals/g mole. In that respect, the effect of small temperature variations on release rates would be minimal.

INDEX